A Research Agenda for Geographies of Slow Violence

Elgar Research Agendas outline the future of research in a given area. Leading scholars are given the space to explore their subject in provocative ways and map out the potential directions of travel. They are relevant but also visionary.

Forward-looking and innovative, Elgar Research Agendas are an essential resource for PhD students, scholars and anybody who wants to be at the forefront of research.

Titles in the series include:

A Research Agenda for Geographies of Slow Violence

Making Social and Environmental Injustice Visible

Edited by

SHANNON O'LEAR

Professor, Department of Geography and Atmospheric Science, and Environmental Studies Program, University of Kansas, USA

Elgar Research Agendas

EE Edward Elgar
PUBLISHING

Cheltenham, UK • Northampton, MA, USA

Published by
Edward Elgar Publishing Limited
The Lypiatts
15 Lansdown Road
Cheltenham
Glos GL50 2JA
UK

Edward Elgar Publishing, Inc.
William Pratt House
9 Dewey Court
Northampton
Massachusetts 01060
USA

Paperback edition 2022

A catalogue record for this book
is available from the British Library

Library of Congress Control Number: 2021935864

This book is available electronically in the **Elgar**online
Geography, Planning and Tourism subject collection
http://dx.doi.org/10.4337/9781788978033

MIX
Paper from
responsible sources
FSC   FSC® C013604
www.fsc.org

ISBN 978 1 78897 802 6 (cased)
ISBN 978 1 78897 803 3 (eBook)
ISBN 978 1 0353 0900 9 (paperback)

Printed and bound by CPI Group (UK) Ltd, Croydon, CR0 4YY

*This book is dedicated to readers who are willing to see violence
and to create a space for justice.*

Contents

Contributors

Daniel Abrahams's research focuses on organizational efforts to address environmental change in places experiencing, or that have recently experienced, acute conflict or disasters. He is currently an American Association of the Advancement of Sciences Science and Technology Policy Fellow at USAID's Office of Global Climate Change working on the Agency's efforts to reduce national-scale greenhouse gas emission through landscape conservation. He is also a Research Associate in Environmental Studies at Colby College.

Sheridan Bartlett works primarily on issues of urban poverty as they affect children in low-income countries, bridging the gap between the concerns of child-focused agencies and the broader development agenda. She is co-editor of the journal *Environment and Urbanization*, International Institute for Environment and Development, London, and an associate at the Children's Environments Research Group at CUNY Graduate Center in New York.

Joseph P. Brewer II is Associate Professor of Environmental Studies at the University of Kansas where he directs the Indigenous Studies Program. His research focuses on working with Indigenous peoples on land stewardship initiatives, relating to natural resources, land tenure, energy sovereignty, and local/regional Indigenous knowledge.

Megan Butler has a PhD in Natural Resources Science and Management from the University of Minnesota. Her dissertation work focused upon community-based forest enterprise governance in the Maya Biosphere Reserve. Her research focuses upon international development practice, human dimensions of natural resource management and community environmental stewardship. She is passionate about creating bridges between research and practice.

Michele E. Commercio is a political scientist at the University of Vermont who specializes in regime transition, ethnic conflict, gender, and Islam in post-Soviet Central Asian states. Her research has been funded by IREX, NCEEER, APSA, USIP, NSF, and UVM. Her work appears in *Central Asian Survey*; *Politics, Groups, and Identities*; *Political Science Quarterly*; *Studies in Comparative International Development*; *Nationalities Papers*; *Problems of Post-Communism*; and *Post-Soviet Affairs*. Her book, *Russian Minority Politics in Post-Soviet Latvia and Kyrgyzstan: The Transformative Power of Informal Networks*, was published with the University of Pennsylvania Press.

Thom Davies is an Assistant Professor in Geography at the University of Nottingham specializing in environmental and political geography. When he has time, he researches social and environmental issues, specializing in toxic geography, environmental justice, and refugee displacement. He has conducted qualitative research with marginalized communities in polluted landscapes including Chernobyl in Ukraine, 'Cancer Alley' in Louisiana, USA, as well as refugee camps in France and Bosnia. Thom has written extensively on environmental issues, including a recent co-edited book with Alice Mah, *Toxic Truths: Environmental Justice and Citizen Science in a Post-Truth Age*, published with Manchester University Press.

Jennifer A. Devine is an Assistant Professor in the Department of Geography at Texas State University. She studies the environmental impacts of drug trafficking, sustainable development, environmental justice movements, tourism and cultural heritage management. In addition to academic publications, she regularly contributes to popular media coverage of these topics. Jennifer holds a MSc degree in Gender and Development from the London School of Economics and Political Science and a doctorate degree in Geography from the University of California, Berkeley.

Aaron H. Gilbreath holds a PhD in Geography from the University of Kansas. His work has focused on the place of drugs in American Society with a particular focus on methamphetamine. He currently lives with his family in Maine, where he works as a statistical consultant in the Department of Academic Technology at Bowdoin College.

Samuel Henkin is a senior researcher at the National Consortium for the Study of Terrorism and Responses to Terrorism (START) at the University of Maryland, College Park. His research is broadly concerned with how violence is exercised strategically to accomplish certain spatial, social, and political ends. Specific research interests include spatial relations of violence, critical theory in political geography, and emotional geographies. His publications vary but all advance greater understandings of everyday violence, and the continuum it forms.

John Paul Henry is a geographer and PhD student at the University of Kansas where he studies geopolitics, surveillance, and violence. His research explores the exercise of power through technological communications and visual representation. Through creative collaboration in photography and film he seeks to amplify silenced voices. In 2020 he was awarded the Social Justice Research Award by the University of Kansas's Research Excellence Initiative for his work concerning Cuban immobility.

Jay T. Johnson is Professor and Associate Chair of Geography and Atmospheric Science at the University of Kansas where he also directs the Center for Indigenous Research, Science, and Technology. His research focuses on Indigenous peoples' cultural survival, particularly in the areas of resource management, political and environmental activism at the national and international levels, and the philosophies and politics of place that underpin the drive for cultural survival.

Hannah L. Legatzke is a PhD student in Natural Resource Sciences and Management at the University of Minnesota. Her dissertation work examines how community-based tourism and tourism to national parks influences residents' livelihoods and attitudes toward conservation in the Maya Biosphere Reserve, Guatemala. Hannah's research interests include tourism sustainability, community-managed forests, and participatory research. Hannah holds a BS in Environmental Sciences and Anthropology from the University of Notre Dame.

Shannon O'Lear is a Professor of Geography with a joint appointment in the Geography & Atmospheric Science Department and the Environmental Studies Program at the University of Kansas. She has written extensively on environmental politics with recent book projects including *Environmental Geopolitics* (2018) and a previous volume with Edward Elgar, *A Research Agenda on Environmental Geopolitics* (2019). She has also written on climate science, science and technology studies, and geopolitics and resource issues in the South Caucasus.

Kelly Overstreet is a doctoral student at the University of Kansas School of Public Affairs and Administration. Kelly holds a Master of Arts in Geography and a Master of Urban Planning from the University of Kansas. Kelly's research asks how professionals can create spaces for public engagement that are compassionate, trauma-informed, and empowering. Her research interests include climate adaptation planning, social justice, compassion and emotions in planning and public engagement.

Laura Aileen Sauls is a Leverhulme Early Career Fellow in the Department of Geography at the University of Sheffield, where she researches movements for Indigenous and community forest rights. Her research primarily takes place in Central America, and she collaborates regularly with the El Salvador-based Regional Research Program on Development and Environment (Fundación PRISMA). Laura earned her PhD in Geography from Clark University, after an MPhil in Development Studies (University of Oxford), and a brief stint in the civil service.

Kimberley Anh Thomas is a human-environment geographer and Assistant Professor in the Department of Geography and Urban Studies at Temple University. She takes a political ecology approach to questions about environmental justice, human vulnerability to hazards, and the multi-scalar politics of resource governance in South and Southeast Asia. Her work has been published in peer-reviewed journals including *Global Environmental Change*; *Annals of the American Association of Geographers*; *Geopolitics*; *Water International*, and *WIREs Climate Change*, and has been supported through funding by the Fulbright Foundation, Council of American Overseas Research Centers, and the Institute for Human Geography.

Ruth Trumble is a PhD candidate in the Department of Geography at the University of Wisconsin-Madison. Her research combines critical geopolitics, disaster studies, and science and technology studies to examine the relationship between disaster preparedness, diverse crises, and international aid in Serbia. Beginning in the 1980s when Serbia was a republic of the Socialist Federal Republic of Yugoslavia and continuing to the present day, her research focuses on the production and maintenance of disaster preparedness expertise amid ongoing economic crises.

Peter Vujakovic is Emeritus Professor of Geography at Canterbury Christ Church University in the UK. Peter has been involved in research into cartography since the 1980s, including work on news media mapping and geopolitics, children's maps, mapping for development education, and disability access mapping. He is former editor (now associate editor) of *The Cartographic Journal* and is an expert contributor to *The Times Comprehensive Atlas of the World*. He is co-editor and contributing author to the *Routledge Handbook of Mapping and Cartography* (2018).

1 Geographies of slow violence: an introduction

Shannon O'Lear

Introduction and overview

Unlike other forms of conflict and direct, physical violence, slow violence is difficult to see and measure. If slow violence includes forms of harm and destruction that are invisible, intangible, indirect, and that unfold over time, how is it possible to investigate this kind of violence? More specifically, how do geographers study slow violence? What kinds of questions do geographers ask about slow violence, how do they inquire into the various workings and impacts of slow violence across a range of themes and theoretical frameworks? What kinds of contributions are these scholars making to our understanding of and response to slow violence?

The objective of this volume is to explore how geographers are applying a range of specializations to study, analyze, and draw attention to forms of harm and violence that have not often been at the forefront of scholarly and public awareness. There is need for greater attention on forms of harm and violence that challenge immediately quantifiable measures of damage, and this attention requires us to consider alternative modes of seeing and interpretation.

Geographers are, in many ways, well-suited to extend familiar modes of inquiry and methods to the study of slow violence. Through long-standing practices of mixed methods research, sensitivities to interactions between people and landscapes, recognition of ways in which simultaneous processes shape multiple spatial scales from the body to the state to the long reach of commodity chains, geographers have a diverse, analytical tool kit for making meaningful investigations into slow violence.

This collection is necessarily incomplete since a comprehensive consideration of slow violence could not be contained in a single volume. Instead, this volume demonstrates how geographers and scholars in related fields approach difficult questions of how to see, understand, and respond to forms of violence that tend to be relatively invisible and incomprehensible within familiar conceptual frameworks, legal systems, and political or social response practices.

Slow violence defined

The roots of scholarly work on slow violence trace back to the work of Johan Galtung. In his now famous paper in *Journal of Peace Research* in 1969, Galtung disrupted the war/peace binary by considering violence not like an on/off switch but as having multiple dimensions such as physical or psychological, intended or unintended, and direct or indirect characteristics. Another dimension of violence that Galtung considered is manifest or latent violence, that is, if violence is immediately observable or not yet visible but likely to emerge in the future. He also considered whether or not harm and suffering are avoidable. If harm and suffering are avoidable but still experienced by individuals or groups, those people are prevented from realizing their full potential. This idea aligns with Galtung's conceptualization of violence. Galtung noted that most violence is thought of as physical and involving a subject, an object, and an action, but there can be harm inflicted when the subject, the object, or both are not immediately involved. If there is no subject actively committing harm directly, that is a case of structural violence:

> We shall refer to the type of violence where there is an actor that commits the violence as *personal* or *direct*, and to violence where there is not such actor as *structural* or *indirect* … There may not be any person who directly harms another person in the structure. The violence is built into the structure and shows up as unequal power and consequently as unequal life chances. (pp. 170–171)

Structural violence is the result of an uneven distribution of resources and access (to education, health care, etc.), but more fundamentally, structural violence results when "the power to decide over the distribution of resources is unevenly distributed" (p. 171). Violence that is built into social structures is also more stable than personal, intentional, direct violence that may be associated with wars and visible, physical conflict.

Twenty years after his introduction of the concept of structural violence, Galtung (1990) followed up his earlier work by expanding it to include the notion of cultural violence:

> By 'cultural violence' we mean those aspects of culture, the symbolic sphere of our existence – exemplified by religion and ideology, language and art, empirical science and formal science (logic, mathematics) – that can be used to justify or legitimize direct or structural violence. Stars, crosses and crescents; flags, anthems and military parades; the ubiquitous portrait of the Leader; inflammatory speeches and posters – all these come to mind. (p. 291)

In his discussion of cultural violence, Galtung observed that, "Cultural violence makes direct and structural violence look, even feel, right – or at least not wrong" (p. 291). Later scholars would refer to this type of violence as resulting from, "the 'soft knife' of routine processes of ordinary oppression" (Kleinman et al., 1997, p. x; cited by Wapner 2014, p. 4). This insidious banality renders such violence tedious to extricate and study yet all the more urgent to understand.

In his 2011 book, *Slow Violence and the Environmentalism of the Poor*, Rob Nixon reinvigorated interest in harm and suffering that results from actions or decisions in indirect ways. His focus was on "the representational challenges and imaginative dilemmas posed not just by imperceptible violence but by imperceptible change whereby violence is decoupled from its original causes by the workings of time" (p. 11). Nixon took Galtung's work a step further by recognizing time as an actor (Kaufman 2014). Nixon also emphasized the challenge and the necessity of generating, "arresting stories, images, and symbols adequate to the pervasive but elusive violence of delayed effects" (p. 3). Yet before those alternative stories and images may be devised in a meaningful way, it is necessary to examine dominant narratives, practices, and geopolitical agendas that enable harmful structures to exist. The chapters in this book consider both of those steps to understanding slow violence.

Geographies of violence

Geographers have contributed considerable thought and work on the topics of war and peace (for overviews see Flint 2005, Mamadouh 2005, Kobayashi 2009, Williams and McConnell 2011, Megoran 2011, and Loyd 2012). Geographers have also challenged the binary of war and peace. They recognize that, "conditions of peace and security are often difficult to disentangle from ongoing wars and other forms of political violence" (Kirsch and Flint 2011, p. 3).

Geographers have recognized how peace-building efforts can serve to rein-force existing, negative power dynamics rather than expand possibilities for social justice (Hammett and Marshall 2017). Geographers study present-day conflict not so much as "new wars" that depart significantly from patterns of past conflicts or as best understood by focusing on state actors. Instead, geographers consider temporal roots and spatial contexts of conflict as well as day-to-day politics involved in navigating and maintaining peace (Agnew 2009). In addition to de-centering the state as the key actor in causing or halting violence, geographers have added meaningful perspective to the study of violence by considering spatialities of violence, critical geopolitics, feminist geopolitics, and aspects of slow violence as it relates to environmental themes.

Geographers' attention to spatial relationships brings an important dimen-sion to the study of violence in all its forms. The volume *Violent Geographies* (Gregory and Pred 2007) widened the aperture on how geographers examine violence and encouraged "an engagement with the *politics* of fear, terror and violence" (p. 6, emphasis in original) and the spaces, both material and imagined, through which those processes operate. In a volume titled *From Above: War, Violence and Verticality*, other geographers consider the signif-icance of a view from above such as from a hill top, hot air balloon, satellite, or military drone, and how verticality enables forms of power and violence (Adey et al. 2013). They note, "As an interface of science, ways of seeing and militarism, there are few perspectives more culpable in their enlistment into practices of war, violence and security than the aerial one" (p. 3). Just as impor-tant as looking at spaces of violence from different vantage points, it is also useful to understand how violence has changed in form and political function in difference places and contexts over time. In *Geographies of Violence: Killing Space, Killing Time* (2017), Marcus Doel traces a history of violence to consider how power has been maintained and challenged in a range of cultural, political, and technological contexts ranging from pre-industrial war to motivations and machines for execution and war making.

In other work on violence, conflict databases provide a means for analyzing and comparing trends in conflict, and they rightly recognize the significance of armed conflict for human well-being and diplomacy. Yet geographers have recognized ways in which these databases capture a limited view of conflict par-ticularly in terms of the spatiality of conflict and associated forms of harm and violence (O'Lear 2005, O'Loughlin et al. 2012). Rather than locating violence in terms of geographic coordinates where "hot" armed conflict is happening, it is useful to see violence as "sitting in places" created through multiple, ongoing processes and political decisions that connect and create violence as a process, not an isolated event (Springer 2011; see also Springer and Le Billon

2016). Geographers have studied border regions not as margins of statehood and globalization, but as viable political units where daily routines contribute to unique forms of conflict and peace (Jones 2012, Korf and Raeymaekers 2013, Brambilla and Jones 2020). Additionally, in some cases, substate conflict associated with political instability may be better understood as having deeper roots in historical inequities of economic power and structural violence where certain demographic groups experience disadvantage (Commercio 2017).

Legal geographers have considered a variety of ways in which law and violence intersect. "Legal geographers," Delaney observes, "take us into the workshops where space, law and (in)justice are the means of the co-production of each other. They show us, often in granular detail, how unjust geographies are made and potentially un-made" (2016, p. 268). By establishing what is accepted as possession and dispossession, Blomley notes that, "there is an intrinsic and consequential geography to law's violence as it relates to private property" (2003, p. 121). He considers how three particular geographic concepts, the frontier, the survey, and the grid, establish and perform spatializations of violence and the "possibility of violence" (p. 130). Indeed, "The violences of law are socially selective" (p. 133). This kind of focus on the "political and ideological significance of space and law, and their representations" reflects a critical legal geography (Blomley and Bakan 1992) that can be helpful in untangling ways in which law and legal definitions contribute to slow violence. In particular, critical legal geography is useful in identifying and analyzing asymmetries of power and injustice. For instance, historical legacies of colonial violence of dispossession of Indigenous land continue to be relevant in questions about the meanings of models of property: "Rather than continuing to privilege dominant Western ideas about property, such as those embodied in the ownership model, there is a need to broaden our view of what property is and what property can be. Recognizing and valuing Indigenous worldviews opens up new ways of thinking about property and geography" (Egan and Place 2013, p. 136). Colonial violence extends beyond the realm of land and resources to perpetuate slow violence "in the intimate, embodied, domestic, micro-scale geographies of Indigenous women and children" (de Leeuw 2016, pp. 14–15). How the spaces and jurisdictions defined by law are mapped and thereby communicated and stabilized are unsettled questions of critical legal cartography (Reiz et al. 2018).

Relatedly, how crime is defined, measured, and controlled can perpetuate the slow violence of racism in the form of mass incarceration, exclusion, and other forms of state organized harm emerging from the criminalization of race (Ward 2014). Taking a feminist perspective on intersections of law and violence, Brickell has argued that, in the case of domestic violence in Cambodia, the rule *of* law is often experienced by already vulnerable populations as an "instrument of power and oppression" through the rule *by* law and "gendered injustice" (2016, p. 184). Other work has considered ways in which violence – in terms of physical and mental harm but also as reflected in one's living conditions – "can break bodies and minds and exert an insidious influence ... the effects are immediate but can also linger, undermining health, trust and capability" (Bartlett 2018, p. 159). Violence in the form of physical aggression, but also in the form of absences of safe home, neighborhood, and community environments for children, contributes to ongoing structural violence that could be disrupted through better policy, planning, and public health and safety efforts. In a different realm, analysis of drone violence and vertical battle space illustrates "how the palimpsest of imperial legality inscribed on the target zones (re)invokes the colonial underpinnings of international law, in order to facilitate and legitimate the deployment of violence 'from above'" (Munro 2014, p. 235). All of these examples demonstrate ways in which law, policy, and the interpretation and mobilization of both can create uneven geographies of violence.

Critical geopolitics and violence

To understand critical geopolitics, it is important first to define geopolitics: "Geopolitics, as the struggle over the control of spaces and places, focuses upon power, or the ability to achieve particular goals in the face of opposition or alternatives" (Flint 2006, p. 28). The control of spaces and places may be most familiar in the arena of states' agendas and international relations, but struggles over the control of spaces and places unfold in and focus on many different spatial scales. These struggles involve a wide range of actors and organizations beyond and within the level of the state. Geopolitics is necessarily about practice and actions taken or not taken. It is also about discourse and how those actions and inactions are justified and explained (Ó Tuathail and Agnew 1992). Geopolitical discourse aims to explain how the world works by constructing geographical knowledge that supports a particular view and agenda. To bring a critical perspective to the study of geopolitical discourse means to ask, "fundamental questions of how power works and might be challenged" (Dalby 1991, p. 266). This kind of approach to understanding the world is in contrast to problem-solving approaches that work within the status quo without questioning why it operates the way it does (see Cox 1984). Simon Dalby's early work on critical geopolitics argued that, "we must not limit our attention to a study of the geography of politics within pregiven, taken-for-granted, commonsense spaces, but investigate the politics of the geographical specification of politics. That is to practice critical geopolitics" (1991, p. 274).

Indeed, critical geopolitics pays attention to "the politics of geographical knowledge and the power of geographical representation" (Dodds 2001, p. 470). It provides an approach to questioning assumptions about familiar categories and narratives about geopolitics (Hyndman 2001). Going beyond simple binaries to recognize "violent continuities between war and peace" brings attention to ways in which militarization and structural violence reinforce each other (Loyd 2009, p. 863). Practices of dispossession and the separation of people from safe public, personal, or political contexts are of concern to geographers who might address these forms of violence by representing "geographies of repossession created by the activists in the networks struggling locally and globally for social justice in all its many forms" (Sparke 2007, p. 347). For instance, gentrification as a form of neighborhood transformation can serve to exclude or marginalize some community members or groups, and this decline in participation and recognition results in "everyday slow violence of cruddy, chronic urban inequality" (Kern 2016, p. 543). Geographies of violence matter for understanding how solidarities of oppressed people may emerge and how "resistance 'takes place'" in particular places and contexts (Inwood et al. 2016, p. 63). In communities affected by slow violence, such as places where toxic pollution has harmful impacts over time, much can be gained by paying attention to the spatial and temporal scales of lived experience. It is important to observe slowly, because slow observation can unsettle assumptions and be an act of resistance aimed at achieving environmental justice (Davies 2018). At any spatial scale, when geopolitical narratives and geographic knowledge are used to promote and legitimize forms of violence, even in the name of security, a critical stance that questions categories and labels, particularly labels of space and time, is valuable (Dalby 2010).

Stability and security at one spatial scale may involve inequitable harm at other, less visible spatial scales. A case in point is the Ganges River treaty that established peaceful terms of water sharing between India and Bangladesh. Although India adheres to the mandated stipulations of water release to Bangladesh, seasonal withholding of water flow inflicts mundane violence in the forms of salinated, undrinkable water supplies, harmful irrigation water, and restricted fish catches experienced locally in parts of Bangladesh (Thomas 2017). Hydraulic infrastructure projects aimed at achieving economic outcomes framed as sustainable development may be successful in the eyes of governments and investors, but these projects may carry impacts of social trauma, structural injustice, and slow violence by communities affected by the construction (Blake and Barney 2018). Other forms of development, such as tourism, also generate structural violence of slowly unfolding harm both for humans and for non-human systems. International tourism development integrates both material and discursive violence through processes of "destructive creation" of inequalities, waste, and branding of imagined spaces (Büscher and Fletcher 2017). Violence and dispossession generated by tourism development can be physical, symbolic, or structural, but could be ameliorated with attention to Indigenous, grassroots, and collective forms of tourism that deter colonial "Othering" so integral to much of international tourism development (Devine and Ojeda 2017). Similarly, international efforts at climate adaptation in the Global South can inflict discursive forms of violence on recipient communities that are (re)produced as vulnerable subjects (Mikulewicz 2020). Through "imaginative countergeographies," local residents can counteract these efforts that effectively exclude them even as international agencies and actors promote values of climate change adaptation efforts. Geographical knowledge may be constructed to perpetuate or hide slow violence not only in terms of state level geostrategy or international objectives, but also in ways that become more visible through feminist geopolitical perspectives.

Feminist geopolitics and violence

Feminist geopolitical perspectives have long offered a valuable means of generating the kinds of alternative stories, images, and symbols called for by Rob Nixon and his work on slow violence. Similar to post-colonial studies and the focus of attention on disempowered, marginalized people, a feminist geopolitical view provides a "lens through which the everyday experiences of the disenfranchised can be made more visible" (Dowler and Sharp 2001, p. 169). Feminist geopolitical perspectives comprise, "an embodied view from which to analyze visceral conceptions of violence, security and mobility" (Hyndman 2004, p. 308). An embodied, rather than objective or state-centered, view is important because while, "the state remains a vital subject of interrogation in relation to security, it obscures fear and violence at other scales, beyond its purview" (p. 308). Rather than studying pursuits of security at the spatial scale of the state, feminist geopolitics tends to consider the scale of the body – an embodied perspective. Bodily and day-to-day experiences of violence are often connected to political agendas and actions happening in more powerful places such as the state level or interstate relations. Taking such a view of violence "from below" enables perspective on ways in which violence is legitimized, criminalized, and resisted through the creation and use of different types of spaces (Fluri 2009, Fluri and Piedalue 2017).

A feminist geopolitical perspective is helpful in discerning forms of violence underlying or entwined with what appear to be peaceful relations (Ross 2011). Similarly, the theory and praxis of antiviolence exposes ways in which the status quo is violent (Loyd 2012, p. 479) and offers a "method that draws connections among multiple forms of violence in order to delegitimize war and other socially sanctioned forms of violence" (Loyd 2012, p. 485). A feminist geopolitical perspective also considers both fear (Pain and Smith 2008) and intimate violence not as a restricted form of oppression, but as spanning the proximate to the distant and involving practices that serve to connect or to disconnect the body to wider geopolitical events (Pain and Staeheli 2014; see also Pain 2015). This kind of perspective asks the question, "How is intimacy" – and associated forms of violence – "wrapped up in national, global and geopolitical processes and strategising, international events, policies and territorial claims, so as to already be a fundamental part of them? ... Which violences receive attention and resources, and from whom? How does their everyday framing as intimate or geopolitical work to sustain them?" (Pain and Staeheli 2014, p. 345). These kinds of questions and insights, possible from a feminist geopolitical perspective, open potential inroads to seeing and responding to slow, banal, and structural violences that might otherwise be invisible or taken for granted.

The very notion of slow violence, as opposed to fast or immediate violence, has been challenged from a feminist geopolitical perspective as representing one side of a limiting binary (Christian and Dowler 2019). From that view, a focus on slow violence overlooks ways in which feminist perspectives have emphasized forms of gendered, racist, and other forms of harm inequitably built into and embodied in day-to-day experience. Indeed, even Galtung (1969) originally recognized that harm and violence involve multiple dimensions that are often co-constitutive of each other. This point is important to bear in mind if we are to understand fully how intertwined political, economic, and cultural structures and practices perpetuate forms of harm both visible and invisible. This volume's attention on slow violence is not exclusive, and many of the chapters recognize how various forms of latent, invisible harm are connected to complex and immediately harmful systems and practices. The focus on slow violence here aims to highlight how we can interpret and understand slow violence not as an alternative to other forms of violence, but both as an often overlooked aspect of normal routines and as an unintended side effect of changes to policy or practice as well as sudden events. What is more, slowing down to pay attention to hard-to-see forms of violence challenges a colonizing notion of time and allows a clearer view of ways in which things are interconnected (Shahjahan 2015).

Slow violence and the environment

Violence is not limited to human systems and structures; it may be intertwined with environmental features. This aspect of slow violence was poignantly illustrated by Michael Watt's seminal book, *Silent Violence* (1983). In this early contribution to geographic work on violence related to environmental features, Watts offered an empirical study of impacts of Nigeria's oil boom. At the same time as the state economy swelled with income from this export commodity, the increased interaction with the global economy contributed to agricultural decline and rural poverty through much of Nigeria. Philippe Le Billon (2001) considered ways in which different types of natural resources are tied to different strategies of power. An edited volume by Nancy Lee Peluso and Michael Watts, *Violent Environments* (2001), brought together a range of perspectives on ways in which environmental features, natural resources, and dimensions of environmental security are intertwined with violence through various tactics, forms of governance, and practices of normalization. Geographic approaches to environmental dimensions of violence tend to avoid analysis focused on resource determinism and have added a nuanced understanding of resource-related conflict as having unique political ecologies (Huber, 2011; see also Peet and Watts 2004) and important geopolitical implications (see, for example, O'Lear and Gray 2006, Selby and Hoffmann 2014, Sneddon 2015, Selby 2018).

Since we often rely on scientific interpretations to shape our understanding of environmental features, it is important to examine how scientific knowledge and practice can normalize some forms of harm and violence. How is scientific practice, attention, and care focused to shape our understanding of human-environment interactions, and how does that focus simultaneously contribute to neglect and harm? As Kathryn Yusoff has argued:

> … violence is a mode of different ontological configuration than care, precisely because it is for the most part banal and indiscriminate, and thus if it is brought to the fore it draws towards the insensible worlds that are beyond the affirmative relations that tend to be prioritized. (2012, p. 590)

Examining how science is done entails examining the "scientific assumptions, discourses, and practices that make nature governable" (Biermann and

Anderson 2017, p. 2), because these processes carry value judgments about how environmental (and sometimes human) features should be managed, protected, or left to die. Scientific and legal practices of establishing categories and lists for sorting and ranking also establish boundaries and embedded values. For instance, paying greater attention to clear and urgent distinctions of 'alien' and 'native' species is the complementary flip side of Nixon's concern with a lack of visibility of slow and structural violence (Lidström et al. 2015). The construction of lists of invasive or alien species in support of narratives of invasion imparts a sense of clarity, boundaries, and certainty around particular species and human interactions with them. Such categorizations become mobile constructs of selective knowledge that, "may preclude diverse ways of knowing and engaging with changing ecologies, landscapes and environments" (Lidström et al. 2015, p. 29).

Temporal and spatial categories may also be usefully disrupted to draw attention to harm and violence that happen in a context of normal environmental policy making. Erin Fitz-Henry's (2020) work on the British Petroleum (BP) Deepwater Horizon oil spill in the Gulf of Mexico in 2014 demonstrates this point. The oil spill was not only extensive, but it was also disaggregated and difficult to track as it was moving in different directions. Corporate and media responses emphasized the importance of returning to normal conditions, but environmental activists challenged this limited and limiting response to the disaster. They argued that this typical response neglected to consider how the oil residue interfered with seasonal and spatial movements of migratory sea mammals and fish into the future as well as the negative, long-term impacts of the oil spill and oil dispersants on all forms of marine life, reproductive cycles, and ecosystem stability. In this work, Fitz-Henry demonstrates the importance of paying attention to slow violence by bringing a deeper, wider, and more integrated understanding of environmental features – and their temporalities and spatialities – to bear on decision-making and disaster management.

Seeing and drawing connections between scientific understanding and decision-making is part of the care of work focused on slow violence. Complex science, such as the state of our knowledge of climate change, can have implications for slow violence when it becomes politically mobilized into selective sound bites that foreclose more transparent policy options (O'Lear 2015, 2016). Other work has considered how the promotion of high-speed, globe-spanning information and communication technologies cannot keep up with the pace of the generation of toxic e-waste (LeBel 2016); "Simply put, planned obsolescence is slow violence" (p. 307). It blinds us to the "unsustainability of the information economy, but makes manifest the gross temporal, economic, and ecological inequities that underpin the ideology and workings of the global information society" (LeBel 2016, p. 308). Again, or perhaps still, in the context of environmental realms, the importance of critical examination of familiar categories and uneven geographies of harm and injustice is abundantly clear.

Overview of chapters

As a research agenda, an aim of this volume is to illustrate different ways that scholars study slow violence through different methodologies and across a range of topics. Each chapter demonstrates a different approach to identifying and interpreting slow violence as one element of a complex context. Thom Davies, an established scholar of geography and slow violence, notes that geographers are well-suited to bring together an awareness of both time and violence through a geographic perspective on how these features quietly unfold in and shape physical spaces. In his chapter titled, "Geography, time, and toxic pollution: slow observation in Louisiana," he demonstrates bearing witness to harm and violence through ethnographic fieldwork focused on the experiences of people living with forms of harm that are difficult to detect. In his chapter, he examines people's experience of living with incremental and gradual harms caused by exposure to industrial pollution. He offers the term and methodology of 'slow observation' to make visible and accessible forms of harm and depleted realities that are perceived only through spending time in a polluted place. People experience pollution and the harm it inflicts on their bodies, environments, and capacities over time. He discusses time as a tool of resistance, and also emphasizes the importance of place.

Ruth Trumble's chapter, "Rhythms of crises: slow violence temporalities at the intersection of landmines and natural hazards," continues with the theme of temporal aspects of slow violence. What happens when an urgent disaster intersects with processes of slow violence? Trumble considers this question in her study of floods in and around Serbia that loosened landmines from their locations and moved them to unknown places. Just as landmines are intentionally placed to alter the navigability of land during a conflict, unexploded ordnance shape post-conflict landscapes in terms of safety and navigation and puts into question whether or not knowledge of landmine placement remains accurate. In 2014, as floodwaters moved through areas known to have been land mined during the armed conflicts of the late 1990s, the location and stability of unexploded ordnance in these places shifted and added a significant yet inconsistent element of danger and risk to flood rescue efforts.

John Paul Henry follows through on Davies's theme of slow observation in his chapter, "Complicating the role of sight: photographic methods and visibility in slow violence research." The question of how to make slow violence visible is at the center of Henry's study of people's lived experience of the toxic landscape of Calvert City, Kentucky. Harm is not necessarily obvious, and Henry emphasizes the role of sight – not the sight of the researcher, but of people who are navigating toxic spaces. This chapter builds on Davies's work, both in this volume and elsewhere (2018) that pays attention to how people see harmful processes unfolding in their living environment, and it offers photographic methods as a means to see the often hidden features of degraded landscapes through the eyes and experience of people who live there. Henry considers the photowalk, the photodrive, and photo elicitation as methodologies to make slow violence visible. Henry demonstrates how these visual methods can be a valuable part of a collaborative, creative process when working with communities that are enduring slowly unfolding, toxic harm generated by corporate neglect and a lack of responsible oversight.

The importance and impact of different ways of seeing are further demonstrated in the chapter "Tourism development as slow violence: dispossession in Guatemala's Maya Biosphere Reserve," contributed by Jennifer A. Devine, Hannah L. Legatzke, Megan Butler, and Laura Aileen Sauls. In this chapter, these scholars draw on their fieldwork in Guatemala's Maya Biosphere Reserve to examine the role of geospatial technologies in processes of dispossession. They consider how geospatial technologies are used as ways of seeing that can contribute to slow violence by rendering human aspects of these places invisible. The authors highlight tensions between, on one hand, grassroots, community forestry practices and small-scale tourism activity, and, on the other hand, government-supported, corporate efforts toward development and commercialized tourism. Combining interviews and field observations, they offer insights into how these technologies contribute to knowledge production that shapes geographical imaginaries and harmful realities of dispossession.

In his chapter, Daniel Abrahams considers climate change and violence in a way that challenges mainstream assumptions about how these two processes are connected. In his chapter, "From violent conflict to slow violence: climate change and post-conflict recovery in Karamoja, Uganda," Abrahams takes an alternative view of climate change and violence that focuses not on immediate, physical conflict but on increasing vulnerability and harm emerging in a place of conflict recovery. In his study of Karamoja, Uganda, Abrahams considers how historical changes to the scope and impacts of cattle raids brought on new forms of conflict in this area, and a national effort at disarmament contributed further to violence and human rights violations. The current efforts at post-conflict recovery are challenged by new vulnerabilities involving changes to land use and land tenure policy intersecting with climatic variability. In the past, Karamoja was the site of severe, violent conflict. That immediate and visible conflict has mostly passed. Now, climate change impacts are contributing to livelihood vulnerability, and negative effects on people's sense of agency and identity, and by these means are contributing to the accumulation of harm and violence that are more difficult to detect and measure.

Kimberley Anh Thomas offers a different view of human-environment interactions and slow violence in her chapter, "Enduring infrastructure." She considers how most attention that is paid to hydropower dams happens around the time and immediate impacts of construction: environmental systems and nearby communities are negatively affected if not destroyed outright; electricity, agricultural water supply, and transportation connectivity tend to increase. Longer-term impacts of water management infrastructure, however, are often overlooked. Thomas draws from her work in Southeast Asia and from work on the social science of infrastructure to consider the infrastructural violence set in motion with the construction of hydropower dams. She considers how these structures, even though they may improve some facets of social life in some places or at some spatial scales, disrupt interactions between social and environmental systems in harmful, depleting ways.

Sheridan Bartlett focuses her chapter specifically on children. In "Slow violence and its multiple implications for children" she considers ways in which children's experience of harm and violence often evolves into slow violence that has a lasting impact on their lives. Not only are children disproportionately vulnerable in risky or turbulent situations, their physical and emotional ability to recover from various forms of harm is limited and may, in fact, contribute to an accumulation of negative impacts over time. Exposure to malnutrition, disease, poverty, environmental degradation, social exclusion, displacement, and instability at community and family levels can amplify and complicate the accumulation of harm and injustice in children's lives.

Joseph P. Brewer II and Jay T. Johnson offer a different view on youth and slow violence with their chapter, "For Indigenous youth: towards caring and compassion, deconstructing the borderlands of reconciliation." Focusing on current-day experiences and representations of Indigenous youth, this chapter considers lingering, harmful manifestations of colonial violence. Generational trauma is perpetuated through tangible and intangible forms of hate, and the authors question if such slow violence is somehow reconcilable. They suggest that the examples of racist behavior toward Indigenous youths that they explore in their chapter can only be addressed through transformative justice when people who are inflicting the harm and hate recognize how they benefit from their own role in dehumanizing practices. These authors offer their own, Indigenous perspective on the potential for justice, not as defined by settler-colonial ideologies based on superiority and difference, but through a deeper, day-to-day recognition of and action within 'natural laws of caring and compassion.'

Michele E. Commercio considers slow violence as experienced by a different demographic in a different geographic context in her chapter, "The infliction of slow violence on first wives in Kyrgyzstan." Commercio travels to Kyrgyzstan, where, during Soviet rule, practices of bride kidnapping and polygyny, the taking of multiple wives, were considered immoral and were punishable crimes. These practices, however, have re-emerged since Kyrgyzstan has gained independence. Commercio considers first wives' experiences with slowly unfolding forms of increased vulnerability for themselves and their children when their husbands take a second wife. Without legal recourse or financial independence, first wives tell stories of the slow violence of deprivation and emotional harm happening in a societal, ideological vacuum that does not protect first wives or their children.

Aaron H. Gilbreath considers how labeling groups of people and associating them with geographies of danger generates another form of slow violence. In his chapter, "When rednecks became meth heads: cultural violence, class anxiety, and the spatial imaginary," Gilbreath examines how imagined geographies are constructed to put distance between "us" and a dangerous "other" in his consideration of the deprecating discourse of "meth heads." He takes a historical look at the emergence of the particular stigmatization of rural, lower-income white populations as criminally inclined "white trash." Gilbreath's investigation of this discursive, cultural violence points to contributing factors including anti-drug campaigns, medical experts, news media, policing agencies, and popular television.

In their chapter, "The slow violence of law and order: governing through crime," Samuel Henkin and Kelly Overstreet offer a different look at the construction of criminality. Henkin and Overstreet locate slow violence in urban governance and dominant perceptions of criminality. Unlike visible forms of violence such as police interventions, practices such as policy decisions or structures of legal systems can perpetuate violence quite behind the scenes. The 'law and order' movement, for instance, with roots in a conservative interest in a rigid crime control system, contributes to uneven but persistent forms of slow violence embedded in the 'making' and 'doing' of urban governance. The authors focus on prosecutorial discretion, the authority of prosecutors to decide whether or not to bring charges, and which charges, and to pursue plea bargains in the face of imprisonment or even the death penalty. They examine the power of prosecutors and how their decisions, far from being value neutral, unevenly shape the lived experience of slow violence in urban spaces.

The collection of chapters is rounded out with Peter Vujakovic's chapter, "Dark cartographies: mapping slow violence." In this rich and substantive chapter, Vujakovic offers a valuable consideration of the role cartography and mapping can play in reinforcing slow violence. Maps communicate and stabilize geographic knowledge and have long been used as tools of oppression and discrimination by elites. Counter-mappings, which depict alternative views and relationships, can serve to challenge the status quo and dominant spatial understanding. Vujakovic discusses how maps – thematic, topographic, as well as combinations of these cartographic forms and the cartographic process itself – may be analyzed as texts or tools to study what they communicate and how they serve to sustain (or challenge) a violent status quo. This chapter offers examples that depict how the assumed objectivity and scientific accuracy of cartography is part of the cultural violence that maps and mapping can serve to reinforce.

Finally, a brief closing chapter offers final thoughts about a research agenda for geographies of slow violence.

Together, the chapters in this volume demonstrate a variety of perspectives, approaches, and methodologies that bring attention to various geographies of slow violence. The chapters consider these forms and patterns of violence as intertwined with other spatial, social, and environmental processes, not as stand-alone features. The contributing authors demonstrate how to make slow violence a visible and integral aspect of research by focusing on particular questions, places, demographics, discourses, practices, and policies. The work captured in these chapters highlights the importance of understanding how slow violence is generated alongside or at times hidden behind other processes that are easier to see and measure. As a whole, this collection offers pieces of a research agenda and invites further contributions and inquiries. One objective of this collection is, indeed, to demonstrate many ways of making the harm of slow violence visible. More importantly, the purpose of this collection is to generate awareness, care, and meaningful practice aimed at alleviating avoidable harm and destabilizing the structures that perpetuate it.

References

Adey, Peter, Mark Whitehead, and Alison J. Williams (eds.) (2013), *From Above: War, Violence and Verticality*, Oxford, UK and New York, NY, USA: Oxford University Press.

Agnew, J. (2009), 'Killing for cause? Geographies of war and peace', *Annals of the Association of American Geographers*, 99 (5), 1054–1059.

Bartlett, Sheridan (2018), *Children and the Geography of Violence: Why Space and Place Matter*, London, UK and New York, NY, USA: Routledge.

Biermann, C. and R. M. Anderson (2017), 'Conservation, biopolitics, and the governance of life and death', *Geography Compass*, 11 (10), e12329.

Blake, D. J. H. and K. Barney (2018), 'Structural injustice, slow violence? The political ecology of a "best practice" hydropower dam in Lao PDR', *Journal of Contemporary Asia*, 48 (5), 808–834.

Blomley, N. (2003), 'Law, property, and the geography of violence: The frontier, the survey, and the grid', *Annals of the Association of American Geographers*, 93 (1), 121–141.

Blomley, N. K. and J. C. Bakan (1992), 'Spacing out: Towards a critical geography of law', *Osgoode Hall Law Journal*, 30 (3), 661–690.

Brambilla, C. and R. Jones (2020), 'Rethinking borders, violence, and conflict: From sovereign power to borderscapes as sites of struggles', *Environment and Planning D: Society and Space*, 38 (2), 287–305.

Brickell, K. (2016), 'Gendered violences and rule of/by law in Cambodia', *Dialogues in Human Geography*, 6 (2), 182–185.

Büscher, Bram and Robert Fletcher (2017), 'Destructive creation: Capital accumulation and the structural violence of tourism', *Journal of Sustainable Tourism*, 25 (5), 651–667.

Christian, J. M. and L. Dowler (2019), 'Slow and fast violence: A feminist critique of binaries', *ACME: An International Journal for Critical Geographies*, 18 (5), 1066–1075.

Commercio, Michele E. (2017), 'Structural violence and horizontal inequalities: Conflict in southern Kyrgyzstan', *Politics, Groups, and Identities*, 6 (4), 764–784.

Cox, Robert W. (1984), 'Social forces, states, and world orders: Beyond international relations theory', in R. B. J. Walker (ed.), *Culture, Ideology, and World Order*, Boulder, CO, USA and London, UK: Westview Press, pp. 258–299.

Dalby, S. (1991), 'Critical geopolitics: Discourse, difference, and dissent', *Environment and Planning D: Society and Space*, 9 (3), 261–283.

Dalby, S. (2010), 'Recontextualising violence, power and nature: The next twenty years of critical geopolitics?', *Political Geography* 29, 280–288.

Davies, T. (2018), 'Toxic space and time: Slow violence, necropolitics, and petrochemical pollution', *Annals of the Association of American Geographers*, 108 (6), 1537–1553.

Delaney, D. (2016), 'Legal geography II: Discerning injustice', *Progress in Human Geography*, 40 (2), 267–274.

de Leeuw, S. (2016), 'Tender grounds: Intimate visceral violence and British Columbia's colonial geographies', *Political Geography*, 52, 14–23.

Devine, Jennifer and Diana Ojeda (2017), 'Violence and dispossession in tourism development: A critical geographical approach', *Journal of Sustainable Tourism*, 25 (5), 605–617.

Dodds, K. (2001), 'Political geography III: Critical geopolitics after ten years', *Progress in Human Geography*, 25 (3), 469–484.

Doel, Marcus A. (2017), *Geographies of Violence: Killing Space, Killing Time*, London, UK: Sage.

Dowler, L. and J. Sharp (2001), 'A feminist geopolitics?', *Space & Polity*, 5 (3), 165–176.

Egan, B. and J. Place (2013), 'Minding the gaps: Property, geography, and indigenous peoples in Canada', *Geoforum*, 44, 129–138.

Fitz-Henry, E. (2020), 'Conjuring the past: Slow violence and the temporalities of environmental rights tribunals', *Geoforum* 108, 259–266.

Flint, Colin (ed.) (2005), *The Geography of War and Peace: From Death Camps to Diplomats*, Oxford, UK: Oxford University Press.

Flint, Colin (2006), *Introduction to Geopolitics*, London, UK and New York, NY, USA: Routledge.

Fluri, J. L. (2009), 'Geopolitics of gender and violence "from below"', *Political Geography*, 28, 259–265.

Fluri, J. L. and A. Piedalue (2017), 'Embodying violence: Critical geographies of gender, race, and culture', *Gender, Place & Culture*, 24 (4), 534–544.

Galtung, J. (1969), 'Violence, peace, and peace research', *Journal of Peace Research*, 6 (3), 167–191.

Galtung, J. (1990), 'Cultural violence', *Journal of Peace Research*, 27 (3), 291–305.

Gregory, Derek and Allan Pred (eds.) (2007), *Violent Geographies: Fear, Terror, and Political Violence*, New York, NY, USA: Routledge.

Hammett, D. and D. Marshall (2017), 'Building peaceful citizens? Nation-building in divided societies', *Space & Polity*, 21 (2), 129–143.

Huber, M. T. (2011), "Enforcing scarcity: Oil, violence, and the making of the market', *Annals of the Association of American Geographers*, 101 (4), 816–826.

Hyndman, J. (2001), 'Towards a feminist geopolitics', *The Canadian Geographer/Le Géographe canadien*, 45 (2), 210–222.

Hyndman, J. (2004), 'Mind the gap: Bridging feminist and political geography through geopolitics', *Political Geography*, 23, 307–322.

Inwood, J., D. Alderman, and M. Barron (2016), 'Addressing structural violence through US reconciliation commissions: The case study of Greensboro, NC and Detroit, MI', *Political Geography*, 52, 57–64.

Jones, R. (2012), 'Spaces of refusal: Rethinking sovereign power and resistance at the border', *Annals of the Association of American Geographers*, 102 (3), 685–699.

Kaufman, A. (2014), 'Thinking beyond direct violence', *International Journal of Middle East Studies*, 46, 441–446.

Kern, L. (2016), 'Rhythms of gentrification: Eventfulness and slow violence in a happening neighborhood', *Cultural Geographies*, 23 (3), 441–457.

Kirsch, S. and C. Flint (eds.) (2011), *Reconstructing Conflict: Integrating War and Post-War Geographies*, Farnham, UK and Burlington, VT, USA: Ashgate Publishing Company.

Kleinman, Arthur, Veena Das, and Margaret Lock (eds.) (1997), *Social Suffering*, Berkeley and Los Angeles, CA, USA: University of California Press.

Kobayashi, A. (2009), 'Geographies of peace and armed conflict: Introduction', *Annals of the Association of American Geographers*, 99 (5), 819–826.

Korf, Benedikt and Timothy Raeymaekers (eds.) (2013), *Violence on the Margins: States, Conflict, and Borderlands*, New York, NY, USA: Palgrave Macmillan.

LeBel, S. (2016), 'Fast machines, slow violence: ICTs, planned obsolescence, and e-waste', *Globalizations*, 13(3), 300–309.

Le Billon, P. (2001), 'The political ecology of war: Natural resources and armed conflicts', *Political Geography* 20 (5), 561–584.

Lidström, S., S. West, T. Katzschner, M. I. Pérez-Ramos, and H. Twidle (2015), 'Invasive narratives and the inverse of slow violence: Alien species in science and society', *Environmental Humanities*, 7, 1–40.

Loyd, J. M. (2009), '"A microscopic insurgent": Militarization, health, and critical geographies of violence', *Annals of the Association of American Geographers*, 99 (5), 863–873.

Loyd, J. M. (2012), 'Geographies of peace and antiviolence', *Geography Compass*, 6 (8), 477–489.

Mamadouh, V. (2005), 'Geography and war, geographers and peace', in Colin Flint (ed.), *The Geography of War and Peace: From Death Camps to Diplomats*, Oxford, UK: Oxford University Press, pp. 22–60.

Megoran, N. (2011), 'War *and* peace? An agenda for peace research and practice in geography', *Political Geography*, 30, 178–189.

Mikulewicz, Michael (2020), 'The discursive politics of adaptation to climate change', *Annals of the American Association of Geographers*, 110 (6), 1807–1830.

Munro, C. A. O. (2014), 'Mapping the vertical battlespace: Towards a legal cartography of aerial sovereignty', *London Review of International Law*, 2 (2), 233–261.

Nixon, Rob (2011), *Slow Violence and the Environmentalism of the Poor*, Cambridge, MA, USA: Harvard University Press.

O'Lear, S. (2005), 'Resource concerns for territorial conflict', *GeoJournal* 64 (4), 297–306.

O'Lear, S. (2015), 'Geopolitics and climate science: The case of the missing embodied carbon', in Shannon O'Lear and Simon Dalby (eds.), *Reframing Climate Change: Constructing Ecological Geopolitics*, New York, NY, USA: Routledge, pp. 100–115.

O'Lear, S. (2016), 'Climate science and slow violence: A view from political geography and STS on mobilizing technoscientific ontologies of climate change', *Political Geography*, 52, 4–13.

O'Lear, S. and A. Gray. (2006), 'Asking the right questions: Environmental conflict in the case of Azerbaijan', *Area*, 38 (4), 390–401.

O'Loughlin, J., F. D. W. Witmer, A. M. Linke, A. Laing, A. Gettelman, and J. Dudhia (2012), 'Climate variability and conflict risk in East Africa, 1990–2009', *Proceedings of the National Academy of Sciences of the United States of America*, 109 (45), 18344–18349.

Ó Tuathail, G. and J. Agnew (1992), 'Geopolitics and discourse: Practical geopolitical reasoning in American foreign policy', *Political Geography* 11 (2), 190–204.

Pain, R. (2015), 'Intimate war', *Political Geography*, 44, 64–73.

Pain, Rachel and Susan J. Smith (eds.) (2008), *Fear: Critical Geopolitics and Everyday Life*, Farnham, UK and Burlington, VT, USA: Ashgate Publishing Company.

Pain, R. and L. Staeheli (2014), 'Introduction: Intimacy-geopolitics and violence', *Area*, 46 (4), 344–360.

Peet, Richard and Michael Watts (eds.) (2004), *Liberation Ecologies: Environment, Development and Social Movements*, 2nd ed., London, UK and New York, NY, USA: Routledge.

Peluso, Nancy L. and Michael Watts (eds.) (2001), *Violent Environments*, Ithaca, NY, USA and London, UK: Cornell University Press.

Reiz, N., S. O'Lear, and D. Tuininga. (2018), 'Exploring a critical legal cartography: Law, practices, and complexities', *Geography Compass*, 12 (5), e12368.

Ross, A. (2011), 'Geographies of war and the putative peace', *Political Geography*, 30, 197–199.

Selby, J. (2018), 'Climate change and the Syrian civil war, Part II: The Jazira's agrarian crisis', *Geoforum*, 101, 260–274.

Selby, J. and C. Hoffmann (2014), 'Rethinking climate change, conflict and security', *Geopolitics*, 19, 747–756.

Shahjahan, R. A. (2015), 'Being "lazy" and slowing down: Toward decolonizing time, our body, and pedagogy', *Educational Philosophy and Theory*, 47 (5), 488–501.

Sneddon, Christopher (2015), *Concrete Revolution: Large Dams, Cold War Geopolitics, and the US Bureau of Reclamation*, Chicago, IL, USA: The University of Chicago Press.

Sparke, M. (2007), 'Geopolitical fears, geoeconomic hopes, and the responsibilities of geography', *Annals of the Association of American Geographers*, 97 (2), 338–349.

Springer, S. (2011), 'Violence sits in places? Cultural practice, neoliberal rationalism, and virulent imaginative geographies', *Political Geography*, 30, 90–98.

Springer, S. and P. Le Billon (2016), 'Violence and space: An introduction to the geographies of violence', *Political Geography*, 52, 1–3.

Thomas, K. A. (2017), 'The "mundane violence" of international water conflicts', *Education About Asia*, 99 (2), 36–41.

Wapner, P. (2014), 'Climate suffering', *Global Environmental Politics*, 14 (2), 1–6.

Ward, G. (2014), 'The slow violence of state organized race crime', *Debating Theoretical Criminology*, 19 (3), 299–314.

Watts, Michael J. (1983), *Silent Violence: Food, Famine & Peasantry in Northern Nigeria*, Berkeley and Los Angeles, CA, USA and London, UK: University of California Press.

Williams, P. and F. McConnell (2011), 'Critical geographies of peace', *Antipode*, 43 (4), 927–931.

Yusoff, K. (2012), 'Aesthetics of loss: Biodiversity, banal violence and biotic subjects', *Transactions of the Institute of British Geographers*, 37 (4), 578–592.

2

Geography, time, and toxic pollution: slow observation in Louisiana

Thom Davies

Introduction

Who has *time* to read a whole academic book? I'm only being slightly facetious when I write this.[1] According to Rob Nixon – whose work on slow violence has animated the pages of this book: 'we live in an age of degraded attention spans' (2011, 13); an epoch of 'intensified distraction' (ibid., 279). In a digitised world that seems to be speeding up, the shortness of our attention is only matched – he suggests – by the slowness of sustained environmental damage. From the incremental brutalities of climate breakdown to the gradual accumulation of pollution in the bodies of the poor; and from the permanent loss of entire species to the slowly unfolding toxic injuries of racial capitalism: the world is replete with delayed catastrophes that are 'slow moving and long in the making' (Nixon 2011, 3). Perhaps the brilliance of Nixon's notion of slow violence comes from its simplicity; a vital characteristic for these hurried times. Turn to page two of his book, *Slow Violence and the Environmentalism of the Poor*, and you will find a summation of his concept neatly put in tweet-length prose. If your copy of *Slow Violence* is borrowed from a university library, this sentence may already be underlined by the pens of previous students:

> By slow violence I mean a violence that occurs gradually and out of sight, a violence of delayed destruction that is dispersed across time and space, an attritional violence that is typically not viewed as violence at all. (Nixon 2011, 2)

The idea of slow violence invites us to expand what might constitute harm, provoking us to look beyond the obvious, the immediate, and the spectacular when researching social injustice. It demands we take seriously strains of violence that have, through the camouflage of time, become unmoored from their original causes. As Nixon (2011, 10) articulates, 'to confront slow violence requires ... that we plot and give figurative shape to formless threats whose fatal repercussions are dispersed across space and time'. Thinking in terms of a research agenda for a *geographies* of slow violence, I suggest our discipline is well placed to take up Nixon's task; of decreasing the distance between cause and effect and articulating the ongoing-ness of temporally scattered damage (see Davies 2019). This is in part because the notion of time and the concept violence is already well-travelled terrain within the annals of geography.

Our spatial discipline, for all its many faults, has a long history of writing about time, stretching back at least half a century (for example, Harvey 1969), thus making slow violence a natural ally in our collective conceptual toolkit. As Merriman (2012, 14) notes, 'different conceptualisations of space and time have more-or-less implicitly or explicitly lain at the heart of geographic debate throughout the discipline's complex, fractured history', and this includes early environmental thought (Vidal de la Blache 1926). Indeed, time has been a central feature of our discipline's work, with 'big name' geographers such as Doreen Massey (1992), Nigel Thrift (1995), and David Harvey (1996) having all written extensively about the intimate connections between time and space, through influential concepts such as 'space-time' and 'time-space compression'.

As well as focusing on time, geographers have also contributed extensively to scholarship on violence (see Gregory 2004; Tyner and Inwood 2014; Doel 2017). It should be recognised, too, that throughout much of Geography's history it has been a discipline *of* violence, with the deep connections between the work of geographers and the technologies of imperialism and war having been rightly exposed and criticised (Driver 1992). It is telling, for example, that one of the academic texts found in Hitler's prison cell when he was writing *Mein Kampf* in 1924 was the book *Political Geography* by German geographer Friedrich Ratzel (Bassin 1987). More recently, however, geography has become a discipline *about* violence, often focusing on the many injustices, conflicts, and inequalities that spatialise the world around us. Recent geographical scholarship has examined the deep interconnections between agonism and space (Springer and Le Billon 2016; Davies 2019) and both 'geographies of violence' and 'geographies of peace' are now important subfields within the discipline at large (see Tyner and Inwood 2014; Springer and Le Billon 2016; Harrowell 2018). They fill the pages of entire books and the contents of entire undergraduate modules. Slow violence therefore comes as a welcome addition to how we think about space and violence.

But what does slow violence bring to the existing conversation? Geographers have already examined violence across a range of fascinating forms and scales, including its links to geopolitics (Shaw 2013), neoliberalism (Springer 2009, 2010, 2011; Tyner 2016), migration (Jones 2016; Davies and Isakjee 2019; Davies et al. 2019; Slack and Martínez 2019), the environment (O' Lear 2004, 2016; Yusoff 2012; Davies 2018, 2019), and armed conflict (Gregory and Pred 2007; Forsyth 2019); but also its relationship to smaller-scale 'everyday terrors' (Pain 2014a), such as the body or the home; with feminist geographers in particular highlighting the intimacies of violence (Bagelman and Wiebe 2017; Mustafa et al. 2019). It is beyond the scope of this chapter to discuss the full gamut of geography's violent oeuvre, but what has bound this scholarship together is how we often rely on the work of social theorists in order to make sense of our violent empirics. Violence continues to prove rich philosophical ground for many social theorists, with scholars as notable as Frantz Fanon (1963), Hannah Arendt (1970), Michel Foucault (1977), Walter Benjamin (1921), and Susan Sontag (2003) – among others – having all discussed violence directly in their work.

Geographical analysis has thus been sharpened by a bewildering array of concepts and theories, from Galtung's (1969, 1990) notions of 'structural' or 'cultural' violence, to Spivak's concept of 'epistemic' (1988) harm, and many more besides. In short, 'slow' violence is not the first time, and will not be the last time, theorists have placed a qualifying adjective in front of the word 'violence' to make a broader point. For example, through reading discussions of violence within geographical scholarship, we learn that it can be 'gendered' (Pain 2014b), 'symbolic' (Bourdieu 1979), 'administrative' (Spade 2015), 'colonial' (Fanon 1963), 'epistemological' (Shiva 1988), 'liberal' (Isakjee et al. 2020), 'insidious' (Anderson 2017), 'banal' (Yusoff 2012), 'infrastructural' (Li 2017), 'silent' (Watts 2013), 'abstract' (Tyner et al. 2014), 'everyday' (Mustafa et al. 2019), 'indirect' (Galtung 1969), or 'inactive' (Davies et al. 2017), among many other incarnations. Each one attempts to push the idea of violence in different directions and illuminate situations of injustice that are woven into the very fabric of society. After all, calling something 'violence' does political work. With such a dizzying display of violent concepts, what *exactly* makes 'slow' violence so special?

For this geographer, slow violence encourages us to enlarge the boundaries of what causes suffering; away from the immediate jolt of cinematic forms of harm, towards incremental brutalities that are predicated on layers of accumulated social, political, economic, and environmental inequity. By delinking our geographical imaginations from the shackles of the present, slow violence asks us to delve into the past to exhume the violent structures of social disparity that saturate the contemporary moment and may yet devastate the future. In this sense slow violence is as much a *structural* concept as a timely one and can be compared to Lauren Berlant's earlier concept of 'slow death' (2007), which suggests we look beyond time-bound 'crisis' thinking and focus upon the ongoing-ness of dispersed harms. Slow violence also resonates with Berlant's notion of the 'ongoing present' (2011, 17), discussed in her book *Cruel Optimism*, as the zone where economic and political structures converge in the geographic here and now. As such, slow violence 'moves us away from the *event* of trauma or catastrophe' (Puar 2017, 11, emphasis added) and instead asks us to consider the wider structural forces that conspire to wear down subaltern populations and spaces over time. Indeed, this structural characteristic of slow violence is, in my view, what gives the concept its *bite*. Nixon makes this connection clear in his homage to Johan Galtung at the start of his book, whose influential concept of 'structural violence' (Galtung 1969) is vital to understanding forms of brutalities that are also slow. You only have to look at the subtitle of Nixon's book '… and the environmentalism of the *poor*' (emphasis added) to note that slow violence is not only about time, it is also about inequality.

Nixon joins a long list of well-cited scholars, such as Judith Butler (2009) and Achille Mbembe (2019), who have argued that not all lives – and not all deaths – are afforded the same political value. As Nixon articulates, 'Causalities of slow violence – human and environmental – are the casualties most likely not to be seen, not to be counted' (Nixon 2011, 13).

As we geographers start to employ and encounter the concept more often throughout our work, we must resist lazy applications of 'slow violence' to each and every aspect of harm that happens to intersect with time. Instead of offering a decaffeinated version of slow violence, we must first ask *who* is being impacted by slow violence; *who* is left untouched by such gradual calamities, and, critically, what are the *politics* that underpins and sustains this violence? As geographers have shown elsewhere, whatever kind of violence is present in society, not all bodies – both human and more-than-human – are rendered equally vulnerable to sacrifice. What role, in other words, does the intersection of race, class, gender, species, sexuality, and disability play in the production and experience of slow violence? Put differently, to draw upon 'slow violence' without attending to its *structural* foundations is a complete impoverishment of the concept itself.

In this chapter, I join others in this edited collection by suggesting that we geographers make room for slow violence within our cache of concepts and ideas. Following previous calls for 'taking time seriously' (Adam 2000, 126),

I suggest we pay attention to the potential of bearing *witness* to the shifting temporalities of violence, and crucially – take seriously the perspectives, experiences, and knowledge-claims of those who are living *with* slow violence. In a discussion of how we might resist slow violence, Nixon calls for a 'different kind of witnessing: of sights unseen' (Nixon 2011, 15). As a literary scholar, he directs his appeals to other creative writers, novelists, and journalists who write about the environment, encouraging them to find new ways of speaking about the gradual but deadly calamities that exist in late modernity, stating that 'imaginative writing can help make the unapparent appear, making it accessible and tangible by humanizing drawn-out threats inaccessible to the immediate senses' (Nixon 2011, 15). This chapter responds to this need, using the tools of a geographer and the approach of ethnography; instead of focusing on the narratives of 'writer-activists' (ibid.), which form the focus of Nixon's book (see Vorbrugg 2019), I will draw instead from ethnographic fieldwork that centres the experiences of people who are actually living *with* the slow horror of pollution. By doing so, I push back against the notion that slow violence is necessarily 'out of sight' (Nixon 2011, 2) to the people it is impacting, or indeed that slow violence is always 'inaccessible' (ibid., 15) to the senses. Instead, I suggest that individuals who live with slow violence are better placed than anyone to incrementally and gradually notice pollution through what I term 'slow observation' (see Davies 2018, 2019).

Before I do this however, and before I take us to the toxic geography of Cancer Alley in Louisiana where this ethnography is based, I want to first discuss the wider role that time is playing in our current enviro-political moment. In doing so, I want to expand on the idea that time is central to how people experience pollution and suggest that it is also vital to how we might act politically in an environmentally altered world. Time, in other words, can be a means of resistance.

Time as a tool of resistance

Across the world today, perhaps more than ever before, time is being mobilised in novel political ways. This is particularly true when it comes to environmental activism. As I write this chapter, Extinction Rebellion protestors have brought central London to a standstill in their calls for the UK government to radically cut carbon emissions. The symbol of the Extinction Rebellion, seen on flags, placards, and even on the tattoos of the most committed, is a simple hourglass within a circle. As stated on the movement's website, the circle represents the planet and the hourglass signifies how '*time is rapidly running out*'.

A generation earlier, a similar symbol of non-violence could be found on the walls of student bedrooms and at countless anti-war protests; the iconic 'peace sign' warned against the borderless violence of sudden nuclear annihilation (Alexis-Martin and Davies 2017). Today, however, the pace of threat has changed, but the end is still irrevocably nigh: spurred on by climate science and an increasing realisation that human activity is threatening the future of the Earth, it is the slow creep of *environmental* violence that is mobilising the masses; the Extinction Rebellion symbol is as much a temporal warning as a geographic one.

A similar time-centred activism can be seen in the work of the *Bulletin of Atomic Scientists* – a non-profit organisation comprised of scientific experts, who also use the metaphor of time in order to make their political point. Founded in the wake of the Second World War by academics who had helped develop the first atomic weapons, this collective uses the symbolism of a ticking 'Doomsday Clock' to communicate contemporary hazards, from impending nuclear Armageddon, to approaching environmental destruction: the closer to midnight that the minute hand gets, the higher the risk of global ecological collapse. Each year throughout the *Bulletin*'s history, the minute hand on the Doomsday Clock has fluctuated, from 'seven minutes to midnight' throughout much of the Cold War, to a much safer '17 minutes to midnight' in 1991, after the disintegration of the Soviet Union. By 2016, however, with Donald Trump entering the White House, fears about climate security meant the apocalyptic clock was brought forward to 'two and a half minutes to midnight'. Three years later in 2019, the Doomsday Clock was set to just *two* minutes to midnight: 'the closest it has ever been to apocalypse' (Mecklin 2019).

The Extinction Rebellion symbol and the ticking Doomsday Clock are not unique in using time as a tool of political awakening. The School Strike for Climate movement, for example, which has been animated by the timely actions of Swedish schoolgirl Greta Thunberg, is the paramount example of politicising temporality: this environmental movement uses time – embodied in the form of *youth* – to make its political point. Having started protesting outside the Swedish Parliament in Stockholm in 2018 at the age of just 15, Thunberg's climate protest soon garnered worldwide attention, and in less than twelve months had grown into a global phenomenon. When she addressed world leaders at a UN summit in Poland in late 2018, the image of a child speaking truth to power became an avatar for a global future threatened by the toxic politics of the past. Leading the way, like a climate-justice version of the Pied Piper of Hamelin, hundreds of thousands of children followed Thunberg's example, walking out of schools across the globe to participate in 'Fridays for Future' protests. Time, once more, was a mobilising force.

With thousands of young people protesting on the streets, the medium became the message: after all, when it comes to the planet's environmental future, who has more at stake – and is less to blame – than children? Speaking to political leaders in Brussels in February 2019, Thunberg said: "If you still say that we are wasting valuable lesson time, then let me remind you that our political leaders have wasted decades through denial and inaction." The School Strike for Climate reached its zenith on a single Friday in late September 2019, when 2 million children in 185 countries walked out of school to demand action on the environment. The week that followed, which coincided with a UN 'Climate Action Summit' in New York, saw a further of 5.6 million people taking part in over 6,000 protests across the planet. The geographic scale of this climate rage was unquestionable, but so too, I suggest, was the galvanising role of *time*. For a generation of young people who had joined Thunberg on the School Strike, the gradual harm of climate change was no longer a potential spectre of the future, it was a toxic and very real inheritance.

Thunberg's climate activism was recognised in 2019 when, at the age of just 17, she was nominated for a Nobel Peace Prize by Norwegian politicians. Although the award that year would go to Ethiopian Prime Minister Abiy Ahmed for his efforts to end border conflict in East Africa, Thunberg's nomination tacitly recognised another form of violence: one that did not reveal itself in the immediacy of bloodshed or the spectacle of the present; some forms of violence – the nomination seemed to suggest – were in Rob Nixon's words: 'slow' (2011). If slow violence represents the dominant threat of the day, in the form of climate chaos or toxic pollution, then this kind of time-based activism from groups such as Extinction Rebellion or the School Strike for Climate, among others, are its revolutionary counterpart. As Nixon suggests, 'to render slow violence visible entails, among other things, redefining *speed*' (Nixon 2011, 13, emphasis added). Through the work of these environmental movements we can see how speed and time can become political. Armed with the persuasive power of time, such environmental movements participate in what Amy Piedalue has called 'slow nonviolence …' (Piedalue 2019, 6), which she describes as efforts that '… work to undo violence in the future as well as in the present, seeking intergenerational, long-term social change that leads to the prevention of violence'. If time has become a mobilising force within global environmental movements, what role does it play for communities who are living at the sharp end of slow environmental crises?

Beyond the whiteness of planetary thinking

What happens when we shift our attention away from the relative safety of Thunberg's home in Sweden, or the occupied streets of London, and focus instead on places where the toxic materiality of the Anthropocene is actually being produced, witnessed, and inhaled? How is *time* being mobilised, experienced, and observed by people who are living *with* the 'peculiar terror of pollution' (Davies 2018, 1549)? People for whom the environment is already breaking down before their eyes, and has been for many years.

As critical race scholars have been arguing for some time, talk of the Anthropocene, climate change, or species extinction 'is configured in a future tense rather than in recognition of the extinctions already undergone by black and indigenous peoples' (Yusoff 2018, 51). Moreover, as the environmental scholar David Schlosberg articulates, 'there is an empty space, or even a negative space, in much Anthropocene writing in relation to justice and environmental justice' (Schlosberg 2019, 54). It must be recognised that the 'belated casualties' (Nixon 2011, 14) of the climate's slow collapse were not, are not, and will not be evenly distributed and experienced. Instead, like other kinds of environmental injustice, they are tethered to the structural inequalities of society, and will be felt most acutely by the poor, the marginalised, and the "epistemically-ignored" (Davies and Mah 2020). What role, therefore, does time play in already-polluted places; communities for whom anthropogenically altered environments are not a forthcoming geological threat but an actually existing present? What if, instead of thinking about slow violence at the global or metaphorical level, we brought place and locality back into the conversation?

Perhaps all this focus on *time* has made us lose sight of *place*? I know what you're thinking – of course a geographer would say this – but hear me out. In thinking planetarily and geologically about environmental problems, it is all too easy to overlook the local – the sticky and thick unevenness of place. In doing so, we risk flattening the very inequalities upon which climate change is built, crushing responsibility for slow violence under the collective temporal weight of 'the age of the human'. While the Anthropocene *is* unfolding globally, and at an unprecedented pace, it also manifests in local and perilously uneven ways, lodging its toxic discard into the ecosystems, landscapes, and lungs that Capital has rendered most vulnerable to sacrifice.

In what remains of this chapter, I will focus on one such locality: the toxic geography of Cancer Alley in Louisiana, which hosts the largest assemblage of petrochemical infrastructure in the Western hemisphere. By introducing and exploring this former plantation landscape and the toxic brutalities it reproduces, I will not only suggest that slow violence is a lived reality for many people there, I will also argue that environmental violence often unfolds at a speed and volume that is entirely visible and noticeable to local inhabitants. In fact, I want to temper the dominant trope in much environmental scholarship and discard studies – a trope repeated in the book *Slow Violence* (Nixon 2011): that pollution is necessarily inaccessible, invisible, or 'out of sight' (Nixon 2011, 2) to the people impacted by it. Much of the argument within Nixon's book rests on the assumption that pollution is 'spectacle deficient' (Nixon 2011, 47). Indeed, what seems to have seduced many geographers who use slow violence is the claim it is 'invisible'. This may be the case. But I want to caution and encourage a further question for critical geographers: invisible to *whom*?

Having conducted research with people living in various toxic geographies, from contaminated refugee camps in France and Bosnia (Dhesi et al. 2018), to the Chernobyl nuclear Exclusion Zone in Ukraine (Davies and Polese 2015), and polluted carbonscapes such as Cancer Alley (Davies 2018), the last thing I would describe these spaces as is lacking in spectacle. Instead, I suggest that it is not the toxic materiality of pollution (i.e. the *substance* of slow violence) that is necessarily invisible to inhabitants of toxic spaces – as local residents testify, it can catch in your throat, sting your eyes, or manifest necropolitically in hospital records or the graveyards of sacrifice zones. Rather, it is the people themselves who are reduced to invisibility; their stories, perspectives, and lives become overlooked and unnoticed, to the point that they are rendered expendable. Slow violence may produce 'staggered and staggeringly discounted casualties' (Nixon 2011, 2), but this process is not removed or 'out of sight' from the people living through it. Rather, part of becoming 'discounted' is having your stories and knowledge-claims ignored (Spivak 1988).

It is at this point that I turn to my ethnographic research in Cancer Alley in Louisiana. Ethnography is itself a slow way of doing research that centres the knowledge claims of local people, and it can act as a foil to faster 'big data' approaches which dominate toxic investigations. As I have written elsewhere (Davies 2019, 6), ethnographic engagement reflects the gradual temporalities of slow violence and 'complements the representational accounts of slow violence offered by Nixon's literary focus' (for a deeper discussion of ethnography and slow violence, see Vorbrugg 2019).

From sugar plantations to chemical plants

'Everyone round here knows someone with cancer …' explained Caitlynn, who has lived in the small rural town of St. James her whole life. Her wooden house backs onto land once owned by the sugarcane farmers who called her ancestors property. Today, the view from her veranda is less bucolic than it once was, with nine large oil-storage tanks rising up from the former plantation next door; just some of over 100 tanks that store petrochemicals, and occasionally leak them, within a mile of her home. '… It's the air' she explained, pointing to the tanks, before listing some of the 12 petrochemical companies that had moved into her majority African American neighbourhood since she was a child. 'We know we're gonna die, just here you die a little quicker!' joked another resident, who lives in the same riverside town. St. James is not exceptional in hosting such a dense assemblage of toxic facilities. Nor are Caitlynn and her neighbours alone in highlighting the risks of pollution in this region: in 2015, a National Air Toxics Assessment conducted by the Environmental Protection Agency (EPA) declared that another predominantly African American town a few miles downstream from St. James, called Reserve, had the most polluted air in the USA. The risk of cancer here, and in other towns across Cancer Alley, is up to 50 times the national average (EPA 2015), creating a situation described in *The Guardian* as 'a slow-motion environmental and public health disaster' (Zanolli 2019).

By night, the chemical plants and refineries in Cancer Alley look like cities. Floodlit and fuming, their flares burn like candles above the cypress trees and swamps that are iconic to this part of Louisiana. By day, the flat sugarcane fields of the lower Mississippi roll out towards the horizon, where they meet the forested bayou beyond. As you drive along the concrete highways and interstates you can see great ships the size of buildings framed above the levee. Miles inland, they float slowly towards deep-water docks in places like St. James, or back towards the Gulf of Mexico, packed with a petrochemical cargo that has given this region its corporate moniker: 'the chemical corridor'. Communities here are surrounded by the infrastructural innards of the oil industry, comprised of fractionating columns, catalytic crackers, and visbreakers; the whole toxic geography criss-crossed by pipelines. As one resident described during a focus group: 'We're penned in on all sides!' This metallic assemblage helps fractionate crude oil into a cocktail of chemical derivatives, including benzene, ethylene, high octane gasoline, diesel, sulphur, and plastic pellets, among many other mundane, useful, and toxic future-artefacts of the Anthropocene.

Like the ships that carry their cargo around the world, it is not difficult to connect this sublime 'oil assemblage' (Watts 2015, 236) to a *global* form of slow violence. With Louisiana hosting one-tenth of US natural gas reserves and around 20 per cent of total oil reserves (ibid.), it is doubtless that some portion of the microplastic flotsam floating in the world's oceans today (and tomorrow) would have been birthed in Cancer Alley, near Caitlynn's home. Such is the magnitude of chemical production between New Orleans and Baton Rouge that individual River Parishes in Cancer Alley have higher annual carbon footprints than entire sub-Saharan countries. Yet however 'global' we wish to think about the environmental impacts of Cancer Alley, we must not overlook or discount the communities who actually live here. Indeed, this petroscape not only produces the by-products of the oil industry, it also releases other slower and less immediate residues of racial capitalism; residues that are unevenly embodied and experienced by frontline communities that themselves actively describe this pollution as a form of violence.

Like many forms of slow violence, the gradual suffering that people experience in Cancer Alley is built upon layers of structural inequality (Galtung 1969). In the middle of the 20th century, petrochemical companies purchased large tracks of riverside property – former plantations – from white landowners who inherited this prime real estate from their slave-owning forefathers (Davies 2018; Rapson 2018). The largest petrochemical company in Louisiana, Dow Chemical, even proudly proclaims this post-colonial history on its website: 'Dow's operations in Louisiana began in 1956 with the purchase of four plantations in Iberville and West Baton Rouge parishes.' When toxic industry moved in, the white population could afford to relocate away from the nascent chemical plants and plastic factories. Their African American neighbours meanwhile, who owned smaller plots of land next door purchased after the abolition of slavery, were not so lucky.

To paraphrase Rob Bullard, slow violence always follows 'the path of least resistance' (1990), and in the former plantation geographies of Cancer Alley the path of least resistance has meant toxic pollution is more likely to impact the African American population than their white counterparts. As one resident of St. James explained: 'The way the racism plays is that the majority of the chemical plants, wherever you go in this state, they are by the black communities.' Slow violence here is distinctly racialised, and as in other settler-colonial contexts environmental racism is not just built on 'white privilege' but white supremacy (see Pulido 2015). In her superb essay 'Plantation futures', geographer Katherine McKittrick argues that 'we can trace the past to the present and the present to the past through geography' (2013, 7). In the post-colonial landscape of south Louisiana, we can see this racial topography come alive in noxious, slow, and deadly ways: the plantation geographies of yesterday mapping directly onto the toxic geographies of today (see Davies 2018, 2019). By examining the toxic geographies of Cancer Alley, we can see that the plantation, as a historical, spatial and political technology, can itself be viewed an incubator of slow violence, with toxic pollution being just the latest in a long line of brutalities justified through racial domination. Simply put, the chemical geographies of Cancer Alley are white supremacy in molecular form.

Slow observation: slow violence in high definition

'The number of times in the morning I open the windows and it smells bad', explained one resident, to agreement from others in the room. Indeed, the consequences of this toxic geography are acutely felt by local residents, who described waking up in the middle of the night to the stench of chemical pollution; or the multiple forms of illnesses colloquially attributed to toxic discard – from asthma to skin rashes, and headaches to leukaemia. In interviews, residents also described being disturbed by the ever-present *noise* of industry, as well as seeing cracks slowly appearing around joists and doorframes caused by the pile-driving of new construction or ships hitting against the sides of the nearby levee. This was toxic pollution spelt out in Technicolor; slow violence in high definition, not some invisible spectre that needed outsider scientists or writer-activists to verify or translate: 'A lot of people are coming down with respiratory disease and especially cancer … It's just terrible…' explained another participant, in a statement common throughout the interviews: '… everybody gets sick!'

It wasn't just the 'immediate senses' (Nixon 2011, 15) that allowed residents to witness the ongoing presence of slow violence. Having lived in this landscape for decades – sometimes entire lifetimes – people here have become attuned to pollution not only through smell, noise, and the deaths of family and friends, but also through gradual changes to the environment itself. During my research in St. James I would be taken on mini-toxic tours of people's back yards and around the local neighbourhood. Participants would point out vegetables, leaves, flowers, and fruit that had changed uncannily over the years: fruit trees that no longer bore fruit or had slowly died; or vegetation that was not as green as it once was. 'We used to fish. And eat the fish' explained one elderly resident of St. James who reminisced about spending time with his father by the banks of the Mississippi before industry arrived: 'You go and eat fish today! When you get that fish out the river and you cut that fish open you can see that jelly … That's the oil in there and you can kind of like smell it a bit, too.'

In group discussions, residents would draw on the slow environmental knowledge they had accrued over time, commenting on how certain animals no longer appeared in the environment where they used to, or how birds avoided certain areas. In Rachel Carson's path-leading book *Silent Spring* (1962), which Rob Nixon cites as a key inspiration (2011), she described a dystopian future, where 'the rapidity of change and the speed with which new situations are created follow the impetuous and heedless pace of man rather than the deliberate pace of nature' (Carson 1962, 24). In St. James, local residents who had lived among industry for most of their lives were witnessing the destruction of nature's deliberate pace in real time. Interviewees described how pollution was altering the rhythms and seasonality of certain plants, with some vegetation flowering at odd times, or not at all. Such environmental and 'bodily reasoning' (Shapiro 2015, 368) that people exposed to pollution use to understand environmental risk has a clear temporal dimension, where slow changes to the local environment are used to make sense of uncanny toxic geographies. This ability to gradually notice the accumulation of environmental change over time is what I have termed 'slow observation' (Davies 2018), which can act in harmony with slow violence, and help to counter it. The slow drip of deteriorating environments becomes a barometer for understanding the impacts of slow violence.

Slow observation is tethered to the toxic experiences of environmental justice communities; their deep connection with place and its shifting temporalities. As such, slow observations are predicated on the notion of *being with* a place and time. It is a form of temporal noticing, that echoes the gradual accumulations of pollution spelt out in the horror of what Rob Nixon called 'slow violence' (2011). Slow observations not only produce 'local knowledge' in a *geographical* sense but also in a temporal one; they are attuned to distinct temporalities, amassed and negotiated across years of prosaic and embodied encounters with pollution (Davies 2019). These slow observations produce an incremental stratigraphy of knowledge: fragmentary, drip-fed, uncanny, and temporally situated understandings. Like the School Strike for Climate movement or Extinction Rebellion, time here becomes an active resource that can have political value. Slow observation has become key to how people here understand and communicate slow violence. In a letter published by a local newspaper which called for the government to prevent the construction of a large petrochemical plant in St. James, one resident wrote: 'Our fig and pecan trees kept us well fed and even provided enough for us to sell. Now the land and everything that grows on it is poison.' Slow observations provide a critical mechanism through which environmentally exposed communities can understand persistent environmental pollution, and perhaps even find ways to resist its longue durée.

Conclusion

It is important to recognise that communities along Cancer Alley are not passive to the onslaught of slow violence. To take just one example, in October 2019 local environmental activists, many of whom live in contaminated areas including St. James, took part in a 13-day march from New Orleans to Baton Rouge, to protest against pollution. With placards reading 'St James is full' and 'No jobs on a dead planet', this environmental collective – called 'the Coalition Against *Death Alley*' – emphasised the necropolitical violence of pollution as well as its delayed and latent effects. In speeches held at various key points along Cancer Alley, the protestors accused industry of causing 'environmental genocide' and described how residents 'are being slowly killed'.

In the conclusion to his book, in typically brilliant prose, Nixon asks: 'In an age that increasingly genuflects to the digital divinity of speed, how will environmental activists negotiate the representational challenges of slow violence – a violence that is by definition image weak and demanding on attention spans?' (Nixon 2011, 276–277). My answer, as a geographer and an ethnographer, is that we must listen to people who are living *with* slow violence. It is their experiences that are already full of the images and spectacle ostensibly missing from discussions of slow violence; their knowledge is both 'local' in a geographic sense, but it is also temporal; and we should make time to listen. To do so will help us better understand, articulate, and perhaps even resist the onwards march of slow violence.

In her timely essay 'The temporal gaze', Barbara Adam (2000, 140) concludes that a 'time-sensitive scholarly enterprise is both the challenge and the task that confronts social theory at the beginning of the new millennium'. Over two decades later, the importance of thinking critically about the power of time still remains, and I suggest that we geographers take up 'slow violence' as part of our conceptual repertoire. Geographers must be time-sensitive and think critically about the power of time, but more importantly we must centre the experiences and expertise of those who are living alongside slow violence, their knowledge claims, and their slow observations. Slow violence is not always invisible. It just depends on who is looking.

Acknowledgements

I am entirely indebted to the communities in Cancer Alley who welcomed me into their homes, gardens, and churches to help me understand what it is like to live with petrochemical pollution. This research would not have been possible without their time and generosity, and this chapter is written in solidarity with their environmental justice struggle: St. James is full! I am grateful to Professor Alice Mah (University of Warwick) for giving me the opportunity of working on the Toxic Expertise project, and for her advice and support. This work was funded by the European Research Council (grant number 639583).

Note

1. Geographers, like other academics, are working under increasingly time-pressured conditions. The velocity of our scholarship has changed, bending to the whim of the neoliberal university (Berg et al. 2016). The twofold rush of metrics and precarity has coerced us into producing academic 'outputs' at breakneck speed (Harrowell et al 2018; Clare 2019). An ethics of 'slow scholarship' (Hartman and Darab 2012; Mountz et al. 2015) has been proposed as the antidote, with the axiomatic realisation that 'earth writers' (Springer 2017) need time and space in order to *think*. And yet, for most early career academics, such appeals to 'slow down' feel frustratingly like the privilege of the tenured (Davies et al. 2021). Time, it seems, has taken hold of us and refuses to let go.

References

Adam, B. (2000), 'The temporal gaze: The challenge for social theory in the context of GM food', *The British Journal of Sociology*, **51**(1), 125–142.

Alexis-Martin, B. and T. Davies (2017), 'Towards nuclear geography: Zones, bodies, and communities', *Geography Compass*, **11**(9), 1–13.

Anderson, B. (2017), 'Cultural geography 1: Intensities and forms of power', *Progress in Human Geography*, **41**(4), 501–511.

Arendt, Hannah (1970), *On Violence*, New York: Houghton Mifflin Harcourt.

Bagelman, J. and S. Wiebe (2017), 'Intimacies of global toxins: Exposure & resistance in "Chemical Valley"', *Political Geography*, **60**, 76–85.

Bassin, M. (1987), 'Imperialism and the nation state in Friedrich Ratzel's political geography', *Progress in Human Geography*, **11**(4), 473–495.

Benjamin, W. (1921), 'Critique of violence', in Martin McQuillan (ed.), *Deconstruction: A Reader*, New York: Routledge, pp. 62–70.

Berg, L.D., E.H. Huijbens, and H.G. Larsen (2016), 'Producing anxiety in the neoliberal university', *The Canadian Geographer/Le Géographe canadien*, **60**(2), 168–180.

Berlant, L. (2007), 'Slow death (sovereignty, obesity, lateral agency)', *Critical Inquiry*, **33**(4), 754–780.

Berlant, Lauren (2011), *Cruel Optimism*, London, UK: Duke University Press.

Bourdieu, P. (1979), 'Symbolic power', *Critique of Anthropology*, **4**(13–14), 77–85.

Bullard, R. (1990), *Dumping in Dixie: Race, Class, and Environmental Quality*, Boulder, CO: Westview Press.

Butler, J. (2009), *Frames of War: When is Life Grievable?* London, UK: Verso Books.

Carson, R. (1962), *Silent Spring*, Greenwich, CT: Fawcett Publishers.

Clare, N. (2019), 'Can the failure speak? Militant failure in the academy', *Emotion Space and Society*, **33**, 1–4.

Davies, T. (2018), 'Toxic space and time: Slow violence, necropolitics, and petrochemical pollution', *Annals of the American Association of Geographers*, **108**(6), 1537–1553.

Davies, T. (2019), 'Slow violence and toxic geographies: "Out of sight" to whom?' *Environment and Planning C: Politics and Space*, 1–19.

Davies, T. and A. Isakjee (2019), 'Ruins of Empire: Refugees, race and the postcolonial geographies of European migrant camps', *Geoforum*, **102**, 214–217.

Davies, T. and A. Mah (2020), *Toxic Truths: Environmental Justice and Citizen Science in a Post-truth Age*, Manchester, UK: Manchester University Press, p. 352.

Davies, T. and A. Polese (2015), 'Informality and survival in Ukraine's nuclear landscape: Living with the risks of Chernobyl', *Journal of Eurasian Studies*, **6**(1), 34–45.

Davies, T., A. Isakjee, and S. Dhesi (2017), 'Violent inaction: The necropolitical experience of refugees in Europe', *Antipode*, **49**(5), 1263–1284.

Davies, T., A. Isakjee, and S. Dhesi (2019), 'Informal migrant camps', in Katharyne Mitchell, Reece Jones and Jennifer L. Fluri (eds), *Handbook on Critical Geographies of Migration*, London, UK: Edward Elgar Publishing, pp. 220–231.

Davies, T., T. Disney and E. Harrowell (2021), 'Reclaiming failure in geography: Academic honesty in a neoliberal world', *Emotion, Space and Society*, forthcoming.

Dhesi, S., A. Isakjee, and T. Davies (2018), 'Public health in the Calais refugee camp: environment, health and exclusion', *Critical Public Health*, **28**(2), 140–152.

Doel, Marcus (2017), *Geographies of Violence: Killing Space, Killing Time*, London, UK: Sage.

Dow (2018), *Dow in Louisiana*, accessed 10 March 2018 at https://www.dow.com/en-us/about-dow/locations/louisiana

Driver, F. (1992), 'Geography's empire: Histories of geographical knowledge', *Environment and Planning D: Society and Space*, **10**(1), 23–40.

EPA (2015), *2014 National Air Toxics Assessment*, accessed 24 April 2018 at https://gispub.epa.gov/NATA/

Fanon, Frantz (1963), *Concerning Violence*, London, UK: Penguin.

Forsyth, I. (2019) 'A genealogy of military geographies: Complicities, entanglements, and legacies', *Geography Compass*, **13**(3), 1–11.

Foucault, M. (1977), *Discipline and Punishment: The Birth of the Prison*, New York: Pantheon.

Galtung, J. (1969), 'Violence, peace, and peace research', *Journal of Peace Research*, **6**(3), 167–191.

Galtung J (1990), 'Cultural violence', *Journal of Peace Research*, **27**(3), 291–305.

Gregory, D. (2004), *The Colonial Present: Afghanistan, Palestine, Iraq*, Malden, MA: Blackwell, pp. 1–15.

Gregory, Derek and Allen Pred (eds) (2007), *Violent Geographies: Fear, Terror, and Political Violence*, London, UK: Routledge.

Harrowell, E. (2018), 'Towards a spatialised understanding of reconciliation', *Area*, **50**(2), 240–247.

Harrowell, E., T. Davies, and T. Disney (2018), 'Making space for failure in geographic research', *The Professional Geographer*, **70**(2), 230–238.

Hartman, Y. and S. Darab (2012), 'A call for slow scholarship: A case study on the intensification of academic life and its implications for pedagogy', *Review of Education, Pedagogy, and Cultural Studies*, **34**(1–2), 49–60.

Harvey, David (1969), *Explanation in Geography*, London, UK: Edward Arnold Publishers.

Harvey, David (1996), *Justice, Nature and the Geography of Difference*, Oxford, UK: Blackwell.

Isakjee, A., T. Davies, J. Obradovic-Wochnik, and K. Augustova (2020) 'Liberal violence and the racial borders of the EU', *Antipode*, **52**(6), 1751–1773.

Jones, R. (2016), *Violent Borders: Refugees and the Right to Move*, London: Verso Books.

Li, T. (2017), 'After the land grab: Infrastructural violence and the "Mafia System" in Indonesia's oil palm plantation zones', *Geoforum*, **96**, 328–337.

Massey, D. (1992), 'Politics and space/time', *New Left Review*, **196**, 65–84.

Mbembe, A. (2019), *Necropolitics*, London, UK: Duke University Press.

McKittrick, K. (2013), 'Plantation futures', *Small Axe: A Caribbean Journal of Criticism*, **17**(3), 1–15.

Mecklin, J. (2019) 'A new abnormal: It is still 2 minutes to midnight', *Bulletin of Atomic Scientists website*, accessed 22 August 2019 at https://thebulletin.org/doomsday-clock/current-time/.

Merriman, P. (2012), 'Human geography without time-space', *Transactions of the Institute of British Geographers*, **37**(1), 13–27.

Mountz, A., A. Bonds, B. Mansfield, J. Loyd, J. Hyndman, M. Walton-Roberts, R. Basu, R. Whitson, R. Hawkins, T. Hamilton, and W. Curran (2015), 'For slow scholarship: A feminist politics of resistance through collective action in the neoliberal university', *ACME: An International E-journal for Critical Geographies*, **14**(4), 1235–1259.

Mustafa, D., N. Anwar, and A. Sawas (2019), 'Gender, global terror, and everyday violence in urban Pakistan', *Political Geography*, **69**, 54–64.

Nixon, R. (2011), *Slow Violence and the Environmentalism of the Poor*, Cambridge, MA: Harvard University Press.

O'Lear, S. (2004), 'Resources and conflict in the Caspian Sea', *Geopolitics*, **9**(1), 161–186.

O'Lear, S. (2016), 'Climate science and slow violence: A view from political geography and STS on mobilizing technoscientific ontologies of climate change', *Political Geography*, **52**, 4–13.

Pain, R. (2014a), 'Everyday terrorism: Connecting domestic violence and global terrorism', *Progress in Human Geography*, **38**(4), 531–550.

Pain, R. (2014b), 'Gendered violence: Rotating intimacy', *Area*, **46**(4), 351–353.

Piedalue, A.D. (2019), 'Slow nonviolence: Muslim women resisting the everyday violence of dispossession and marginalization', *Environment and Planning C: Politics and Space*, 1–18.

Puar, Jasbir (2017), *The Right to Maim: Debility, Capacity, Disability*, London, UK: Duke University Press.

Pulido, L. (2015), 'Geographies of race and ethnicity 1: White supremacy vs white privilege in environmental racism research', *Progress in Human Geography*, **39**(6), 809–817.

Rapson, J.K. (2018), 'Refining memory: Sugar, oil and plantation tourism on Louisiana's River Road', *Memory Studies*, 1–15.

Schlosberg, D. (2019), 'Disruption, community, and resilient governance: Environmental justice in the Anthropocene', in Tobias Haller, Thomas Breu, Tine De Moor, Christian Rohr, and Heinzpeter Znoj (eds), *The Commons in a Glocal World: Global Connections and Local Responses*, London, UK: Routledge, pp. 54–71.

Shapiro, N. (2015), 'Attuning to the chemosphere: Domestic formaldehyde, bodily reasoning, and the chemical sublime', *Cultural Anthropology*, **30**(3), 368–393.

Shaw, I.G. (2013), 'Predator empire: The geopolitics of US drone warfare', *Geopolitics*, **18**(3), 536–559.

Shiva, V. (1988), 'Reductionist science as epistemological violence', in Ashis Nandy (ed.), *Science, Hegemony and Violence: A Requiem for Modernity*, Oxford, UK: Oxford University Press, pp. 232–256.

Slack, J. and D.E. Martínez (2019), 'The geography of migrant death: Violence on the US-Mexico border', in Katharyne Mitchell, Reece Jones, and Jennifer L. Fluri (eds), *Handbook on Critical Geographies of Migration*, Cheltenham, UK and Northampton, MA, USA: Edward Elgar Publishing, pp. 142–152.

Sontag, Susan (2003), *Regarding the Pain of Others*, London, UK: Penguin.

Spade, D. (2015), *Normal Life: Administrative Violence, Critical Trans Politics, and the Limits of Law*, Durham, NC, USA: Duke University Press.

Spivak, G. (1988), 'Can the subaltern speak?', in Rosalind Morris (ed.), *Can the Subaltern Speak? Reflections on the History of an Idea*, New York: Columbia University Press, pp. 21–78.

Springer, S. (2009), 'Culture of violence or violent orientalism? Neoliberalisation and imagining the "savage other" in post-transitional Cambodia', *Transactions of the Institute of British Geographers*, **34**(3), 305–319.

Springer, S. (2010), 'Neoliberal discursive formations: On the contours of subjectivation, good governance, and symbolic violence in posttransitional Cambodia', *Environment and Planning D: Society and Space*, **28**(5), 931–950.

Springer, S. (2011), 'Violence sits in places? Cultural practice, neoliberal rationalism, and virulent imaginative geographies', *Political Geography*, **30**(2), 90–98.

Springer, S. (2017), 'Earth writing', *GeoHumanities*, **3**(1), 1–19.

Springer, S. and P. Le Billon (2016), 'Violence and space: An introduction to the geographies of violence', *Political Geography*, **52**, 1–3.

Thrift, N. (1995), 'Classics in human geography revisited: On the determination of social action in space and time', *Progress in Human Geography*, **19**, 528–530.

Tyner, J. (2016), *Violence in Capitalism: Devaluing Life in an Age of Responsibility*, Lincoln: University of Nebraska Press.

Tyner, J. and J. Inwood (2014), 'Violence as fetish: Geography, Marxism, and dialectics', *Progress in Human Geography*, **38**(6), 771–784.

Tyner, J., J. Inwood, and D. Alderman (2014), 'Theorizing violence and the dialectics of landscape memorialization: A case study of Greensboro, North Carolina', *Environment and Planning D: Society and Space*, **32**(5), 902–914.

Vidal de la Blache, Paul (1926), *Principles of Human Geography*, London, UK: Constable.

Vorbrugg, A. (2019), 'Ethnographies of slow violence: Epistemological alliances in fieldwork and narrating ruins', *Environment and Planning C: Politics and Space*, 1–16.

Watts, Michael (2013), *Silent Violence: Food, Famine, and Peasantry in Northern Nigeria*, Vol. 15, Athens, GA, USA: University of Georgia Press.

Watts, M. (2015) 'Securing oil: Frontiers, risk, and spaces of accumulated insecurity', in Hannah Appel, Arthur Mason, and Michael Watts (eds), *Subterranean Estates: Life Worlds of Oil and Gas*, Ithaca, NY, USA: Cornell University Press, pp. 211–236.

Yusoff, K. (2012) 'Aesthetics of loss: Biodiversity, banal violence and biotic subjects', *Transactions of the Institute of British Geographers*, 37(4), 578–592.

Yusoff, Katheryn (2018), *A Billion Black Anthropocenes or None*, London, UK: University of Minnesota Press.

Zanolli, L. (2019), '"If there's a spill, it's a disaster": Living next to a giant lake of radio-active waste', *The Guardian*, accessed 7 November 2019 at https://www.theguardian.com/us-news/2019/nov/06/louisiana-st-james-parish-lake-radioactive-industrial-waste-cancer-town-pollution-mosaic

3 Rhythms of crises: slow violence temporalities at the intersection of landmines and natural hazards

Ruth Trumble

Introduction

Slow violences do not necessarily have steady speeds. Instead, their speeds can change when they intersect with other events and emergencies creating "rhythms of crises." The phrase "rhythms of crises" provides one entry point into discussing the varied and, often, unpredictable temporalities of crises that contribute to the making of a disaster. More than five years after the May 2014 floods in southeastern Europe, the landscape remains contaminated with landmines in unknown locations, which increases the threat of harm to residents and to refugees who pass through to enter the European Union (EU). Focusing on landmines extends our understanding of the temporalities of harm as they do not harm with an even, predictable temporality. Here I discuss the relationship between landmines and the May 2014 floods to examine the multiple temporalities at play when the inconsistent and unpredictable speeds of unexploded ordnances (UXOs) intersect with an acute natural hazard event.

Slow violence alerts us to landscapes of precarity that are not easily seen by the discerning eye. It is a violence of "delayed destruction that is dispersed across time and space" (Nixon 2011: 2). In the case of landmines, the violence wrought by them is detached from the original cause by geopolitical and environmental changes in the landscape. After a conflict, often the population at risk is not the same as the population for whom the mines were intended. Therefore, the harms from landmines are also dispersed across groups in addition to "time and geography" (ibid.). Similar to landmines, the damage of natural hazard events can be long-lasting (Adams 2013; Bankoff 2007), and humans have even less control over where and how natural hazard events occur.

Slow violence is useful as a jumping-off point to think through how uneven temporalities of violence intersect during and after an emergency event. Most scholarship that engages with slow violence discusses environmentalism, environmental inequalities, and climate change (e.g. Davies 2018; O'Lear 2016). Like slow violence, natural hazard events exacerbate the violences of racism, pollution, and economic inequality (e.g. Adams 2013; Anderson et al. 2019; Davies 2018; Pulido 2000), further increasing the vulnerability of populations (Cannon 1994; Wisner and Luce 1993). For example, Hurricane Katrina displayed how slow violences of racism, income inequality, and inefficient infrastructure maintenance compound the effects of an emergency natural hazard event (Bakker 2005; Katz 2008; Smith 2006). Thinking about the intersections of crises in one space as "rhythms of crises" complements Jenna Marie Christian and Lorraine Dowler's call to push beyond the fast vs. slow binaries of slow violence (2019) and scholarship that uses rhythms to address the diverse temporalities of disasters (e.g. Norbert et al. 2017). This chapter uses rhythms to consider the multiple, often-unpredictable temporalities at play when an emergency event intersects with certain types of slow violence.

Unlike instances of pollution where the very cells of the body are harmed daily and with consistency (see Davies 2018), landmines present a threat to bodies via two different temporalities: the chronic lack of access to land due to fear of the mines and the acute explosions of the mines when they are triggered. As the landmines stay in the area in which they were placed during conflict, the geographies of that area continue to shift as political, economic, and social changes occur post-conflict. These UXOs materially demonstrate the long-term violences that linger after conflict ends and the staccato, unpredictable temporalities of a certain kind of slow violence.

Within the literature on landmines there is a small but growing body of scholarship on the relationship between landmines and the 'natural' environment. For example, during winter in Korea's DMZ, landmines freeze in the swampy landscape, which renders them useless until spring (Kim 2016). Some scholars note that landmines can serve as a driver of nature conservation as they deter humans and thus provide an unencumbered space for other species to thrive (Kim 2016; Schwartzstein 2014; Smallwood and Lookingbill 2019). However, often residents put their lives at risk by venturing into areas contaminated with mines to forage for higher-quality food or lumber to sell in cities (Henig 2012, 2019). I extend this scholarship on landmines and the environment by demonstrating how the different temporalities of flood emergency and the slow violence of UXOs interact to reshape a landscape.

Through a discussion of the May 2014 floods in three countries of the former Socialist Federal Republic of Yugoslavia (SFRY)—Bosnia-Herzegovina (BiH), Croatia, and Serbia—I explore how the interaction between the slow violence of landmines and a flood event reshapes precarious landscapes. Each country's recent history since the dissolution of the SFRY in 1990s shapes its relationship with the EU and with the other neighboring countries. This relationship also plays a role in how each country is framed as being a part of Europe or something "not-quite-Europe" (Baker 2018). These three countries are in close physical proximity to the EU, and Croatia gained EU membership in 2013 while the other two are in accession. Further, their physical proximity to the EU is an important factor in why Serbia, and now BiH and Croatia, became important transit countries for refugees to reach the EU.

To articulate the intersections of danger that occurred during the May 2014 floods and in its aftermath, I draw on semi-structured interviews conducted from 2015–2016 with demining professionals in BiH and Serbia; flood survivors of Obrenovac, Serbia; and aid workers in Obrenovac and in Preševo, Serbia. I also use news media and reports about all three countries and field-notes from my visits to flood-affected sites and refugee camps in Serbia. The anonymous interviews occurred in participants' offices or in a public location of their choosing. Interviews lasted from one to four hours. All participants were fluent in English and preferred that we conduct the interview in English.

I first describe UXOs, specifically landmines, in relation to slow violence and briefly discuss their role in southeastern Europe. Next, I discuss the May 2014 floods and how the floods reshaped the spatialities of the mines. The known presence of landmines and their unknown locations influenced emergency response throughout BiH, Croatia, and Serbia and then recovery from the floods. I then consider how exploring the intersection of natural hazard events and the slow violence of landmines extends understandings of the diverse temporalities (or rhythms) of harm. I conclude by stating how this exploration complements the ongoing scholarship on slow violence and refugees and call for future scholarship to explore the threat to bodies created when multiple crises interact.

The devil's gardens, pašteta, and yellow killers

The "devil's garden" is one of many names for an area of land embedded with explosive weapons of war (Monin and Gallimore 2011). Bosnians refer to the PMA-2, an anti-personnel (AP) mine widely used in the war in Bosnia as *pašteta* because it resembles a can of pâté widely available in stores (Henig 2019). This weapon was often used by combatants in the conflicts between Serbia, BiH, and Croatia that occurred in BiH and Croatia from 1991–1994 during the break-up of the SFRY. These were found on Serbia's western border with Croatia until 2009 (*Landmine and Cluster Munitions Monitor* 2012). Today, Serbia continues to have AP mines and cluster munition remnants in the south and along its border with Kosovo, which remain from the Kosovo conflicts in the late 1990s. "Yellow killer" or *žuta ubica* is the local name for the bright yellow cluster munition, BLU-97, that was the most common explosive weapon used during the NATO 1999 bombing in what are today Serbia and Montenegro (Norwegian Peoples Aid 2008). These names are among many for weapons of war and the landscapes they wrought given by those who must live with their danger during and after conflict. After a conflict ends, landmines and other forms of UXOs remain wherever combatants planted them—embedded, for example, in the ground, in homes, and in trees.

These entrenched weapons produce a violence that is inconsistent yet long term. The direct violence of the mines does not seep into the body in the same way that air pollution does (see Davies 2018). Instead, landmines harm bodies directly in punctured moments of explosion. The indirect violence of mines occurs when knowledge of mine locations and fear of unknown, potential locations causes people to shift their mobilities. Those who do not know the landscape, such as refugees, are the most at risk to be harmed by a mine because they lack a knowledge of mine locations and an intimate history of the place.

Landmine is a term for explosive weapons that combatants place in the ground according to the defense needs of a state or population. The term includes anti-tank mines which are designed to destroy vehicles, AP mines that are devised to harm individuals, and bomblets of cluster munitions that do not explode upon impact with the ground (Bolton 2010; Nixon 2007). AP mines are the main type of mine referenced here, yet the area flooded across BiH, Croatia, and Serbia also contained anti-tank mines and bomblets from cluster munitions in the ground. AP mines create a particular type of havoc as they are designed to harm a population by maiming, rather than killing, an individual (ibid.). When the individual is injured, it slows down the opposing population, for one injured person requires far more assistance, aid, and time than a dead one.

The international community now widely considers AP mines a menace to be exterminated (ICBL 2019). Globally, from 1999–2017 unexploded remnants of war caused 120,000 reported casualties (Landmine Monitor 2018). However, the actual number of casualties is higher because people may not report land-mine injuries or deaths due to fear of political repercussions or doubt in inef-ficient institutions (Kim 2016; Norwegian Peoples Aid 2008). Other types of weapons such as "smart" bombs disperse hundreds of thousands of bomblets. Some bomblets do not explode upon impact and, thus, become landmines. Regardless, they are considered by states to be an improvement over AP mines (Anderson 2000; Nixon 2007; Norwegian Peoples Aid 2008; Tyner 2010).

At the 1997 Ottawa Convention, many countries agreed to discontinue the manufacturing of AP mines with the goal of forbidding the "use, stockpiling, production, and transfer of anti-personnel mines and on their destruction." The Mine Ban Treaty, as it is otherwise called, was implemented globally in 1999 (ICBL 2019). Still, some countries, including the United States, refused to sign the treaty. The US cites the Korean Demilitarized Zone (DMZ) as the reason that it must continue to manufacture these weapons (Kim 2016). Most recently in the US, the Trump administration has removed the 2014 ban that prevented landmines from being used outside of the DMZ (Egel 2020).

The work of mines in conflict and after

During a conflict, mines deter opposing forces, and if the lines of combat do not shift, the mines work in favor of those who put the mines in the ground. The security that landmines provide the defending party outweighs the poten-tial risks that the mines hold. However, once placed in the ground, they cannot easily be moved or disarmed. When the battle lines shift or the conflict ends, those who planted the mines must remember where the mines are located or risk the same fate of explosion. Thus, mines serve a dual role by posing a risk for opposing forces (or those unaware) and as an apparatus of security for those who placed the mines. Yet if the lines of the battle change or if the knowledge of mine placement disappears, then the armed landmines remain a threat for all.

In the aftermath of conflict, landmines reshape a landscape into one with unpredictable hazards. The harm caused by their placement continues over time and across space as their existence forces people to avoid the use of contaminated land. With the conflict over, the mines remake the landscape in unforeseen ways (Unruh et al. 2003), and it is only through the specified knowledge of UXO locations that harm can be avoided. While removing the mines completely is the surest way to confirm the safety of those in the landscape, a person can safely navigate the landscape when they know how to identify a mine or the mine's location. Building this awareness is one aspect of demining.

Demining refers to a range of activities that include mapping and marking suspected hazardous areas (SHAs), documenting the clearance of the land, and physically clearing the land of UXOs. The physical removal of mines is the most expensive part (World Bank et al. 2014a; Interview A 2015; Interview B 2016), and the political economy of demining delays the removal of all UXOs from the ground (Bolton 2010). While AP mines are manufactured for the cost of $1–3, it can cost up to $300 to disarm the mine once it is in the ground and armed. Due to this political economy of UXOs, states often leave them in the ground in less populated areas with clear signs marking their locations (ibid.). The terrain also determines what type of demining tools are used for the technical clearing of land. Often a non-technical survey is completed first, using available data such as previous mine locations, or, in the case of a flood, direction of water flow. This "desk survey" is frequently considered the first step in assessing mine contaminated areas (Interview A 2015, Interview C 2015; Interview D 2016; UNDP (HR RNA) 2015).

Disarming UXOs depends on political and economic factors (Bolton 2010; Interview C 2015; Nixon 2007) and also on environmental practicalities. For example, in extremely mountainous regions, handheld metal detectors are the best tools to use as vehicles cannot be used (Interview A 2015; Interview B 2016; Interview C 2015). In areas such as protected forests, experts may recommend against using tractors and dogs so as to not harm the vegetation (UNDP (HR RNA) 2015). Consequently, the demining process in protected areas occurs more slowly. These environmental factors are context dependent and can further slow the process by increasing demining costs. In sum, AP mines and other UXOs are inexpensive to manufacture, easy to arm, and very difficult to remove.

Pollution of landscapes and futures

Landmines and other remnants of war can pollute by secreting toxins into the ground just as waste from a factory (Henig 2012, 2019; Kim 2016; Nixon 2007, 2011). Yet, these mines do not disperse pollution into the environment in the same way as a factory or nuclear reactor. Instead, the mines threaten nearby populations through their armed presence in the ground. For example, the conflicts in the former SFRY throughout the 1990s left toxins and UXOs in the ground and increased the economic, social, and political vulnerabilities of residents (Crowther 2008; Henig 2019; UNDAC 2014). While, AP mines can pollute the ground as they decay, Yugoslav-made AP mines and their trip wires have a plastic coating that delays deterioration (Crowther 2008). The plastic impedes the excretion of toxic materials, but it also ensures that the UXOs will remain in working condition longer than those without the plastic coating. For example, during the war between BiH, Croatia, and Serbia, many mines were placed in the Sava River Basin, which was a strategic point for all three countries (Stec et al. 2011). The Sava River is a natural border between BiH and Croatia, and it continues east into Serbia where it meets the Danube River at Belgrade. Some mines did explode and added heavy metals into the water and soil of the Sava (Miko et al. 1995, in Stec et al. 2011). However, due to the plastic coating, some mines endured, slowly destabilizing and contaminating the land. The passage of time increases the uncertainty of if, when, and how the mines will explode (Bolton 2010; Henig 2012).

In areas with or suspected of having UXOs, children can no longer play, farmers can no longer farm, and life must be lived around these places suspected of hazard (Crowther 2008). In this way, populations need not experience the explosion of a UXO to be harmed by its existence in the ground (Bolton 2010; Nixon 2007, 2011; Tyner 2010). The damage of a landmine occurs in how it reshapes space, forcing people to not use land due to well-founded fear of harm. Thus, mines pollute the lives of those who live nearby as well as deteriorating into the land slowly over time. The mines no longer threaten a specific enemy but instead they deter the mobilities of aid workers, civilians, and refugees. In some cases, landmine contamination creates new and danger-ous livelihoods. Residents often undergo training to work as deminers. Some Bosnians enter areas known to have landmines and other UXOs to harvest timber, because that timber is of high quality as it is protected from human development (Henig 2019). Thus, the legacies of the violent conflicts that were a part of the dissolution of the former SFRY continue.

"An unparalleled disaster": the May 2014 floods

Disasters highlight the moment and place in which a human population does not have the capacity to deal with an extreme natural or human-instigated event (O'Keefe et al. 1976). In these instances, emergency situations snowball beyond the capacities of the affected populations. Therefore, disasters cannot exist without interaction with humans (Cannon 1994; O'Keefe et al. 1976;

Smith 2006). It is at the point of interaction with humans that natural hazard events become disasters and, simultaneously, geopolitical events (Field and Kelman 2018; Hyndman 2011; Kelman 2007; Pelling and Dill 2008).

In May 2014, southeastern Europe experienced the worst flooding in its recorded history as an extratropical cyclone named Tamara came from the Mediterranean Sea and barreled east. BiH was the most affected, followed by Serbia, and then Croatia. Three months' worth of rain fell in three days, and in some places—such as Tuzla, BiH—half a year's worth of rain fell in five days (World Bank et al. 2014a; NASA 2014). The Sava River, between all three countries, was unable to absorb the unprecedented amount of water. The excess water spread throughout its entire watershed, flooding the small creeks and streams, which then flooded villages (NASA 2014; UNDAC 2014). As the rivers broke their banks, states mandated evacuations of more than 60,000 people across the region. The cost of damages was about €2 billion in BiH, €1.5 billion in Serbia, and €40 million in Croatia (UNDP 2015; World Bank et al. 2014a, 2014b). Over 2.6 million people were affected by the flood across the three countries (WHO 2014). In BiH, some of the displaced residents had only recently returned after fleeing the civil war of 1992–1995. Throughout all three countries, people living in small towns and villages within the watershed of the Sava River were the most affected by the floods.

The floods triggered 2,000 landslides throughout the three countries, of which BiH experienced 1,200. Additionally, the majority of the UXOs were in BiH, which remains the most mined country in Europe (Norwegian Peoples Aid 2014). The floods moved many signs that marked UXO locations, and the landslides moved or covered many of the 122,000 remaining landmines in BiH (World Bank et al. 2014a; SMAC 2014). There was popular concern that flood waters would move the lightweight plastic (*pašteta*) mines (Kakissis 2014). The fear of the unpredictability of their explosive potential was further cemented in the minds of the residents and responders when a mine exploded in the Brčko district of BiH though no one was injured (Matic and Smajilhodzic 2014). However, water only moved the mines that were previously in a riverbed (Norwegian Peoples Aid 2014). Landslides were the greater cause for alarm.

The landslides halted transportation, threatened infrastructure, and buried landmines (UNDAC 2014; WHO 2014). By 2014, Serbia had mostly cleared the landmines from the early 1990s' conflicts along its border with BiH and Croatia. Croatia had also cleared many of its contaminated areas; however, certain less-populated areas, such as those along the border with Serbia, remained contaminated (UNDP 2015). Landslides caused a Croatian mine-field along the border with Serbia to shift, raising concerns that UXOs from the

Croatian side had entered Serbia (Interview A 2015; SMAC 2014). The newly unknown whereabouts of landmines posed a challenge to flood response and recovery in all three countries (World Bank et al. 2014a; SMAC 2014).

The floods obscured the once-known locations of the mines. The dangerous potentiality and unknowability of the locations of landmines shaped the entire response and recovery effort. All the previous efforts at managing the weapons embedded in the ground—marking and mapping them—had also been washed away in the flood. Prior to the floods, the urgency to demine had been tampered by the high cost of demining (Bolton 2010; Clayfield 2017). The flood renewed the urgency among local, state, and international actors to survey the inundated land across the three countries for UXOs, to mark the ones that were found, and to manage the slow violence of the mines once again (Interview A 2015; UNDAC 2014). When a natural hazard event intersects with slow violence, both can exacerbate the intensity of the other. The different temporalities came together when the rapid flood waters shifted the land and possibly the already-unstable mines. The case of Sapna demonstrates these intersections of slow violence and emergency.

The town of Sapna, in the Tuzla Canton of BiH, is a specific example of how the multiple temporalities of danger brought together by the UXOs and the flood waters exacerbated the disaster and harmed residents. Prior to the floods the marked mines sat latent, away from populated areas in the town. Clear signage reduced the risk that someone would trigger them. Yet, in the first few weeks after the flood, a demining team found three AP mines that blocked access to the town's water supply. Due to the physical geography of the area, the more efficient larger vehicles could not go over the hills to find mines, and the demining team had to use metal detectors (Chick 2014). Thus, the demining team had to use metal detectors. However, because metal detectors can only detect ordnances up to a foot and a half into land, the mud had to be stripped in small layers to ensure safety in the demining process. This time-consuming process of carefully stripping away the mud added another economic strain to Sapna residents, as they had to buy bottled water during the tedious process of demining and flood clean-up (ibid.). Therefore, disarming the mines that most immediately threatened residents and recovery workers became an emergency priority, for only when the mines were disarmed could flood recovery begin.

In Sapna and beyond, the mines became part of the flood emergency. Finding and marking where the mines had moved, if they had shifted at all, became of paramount importance. The temporalities of the mines shifted as those buried by landslides remained hidden and difficult to trigger, while other mines were brought via water and landslides into areas of daily social life. BiH, Croatia, and Serbia worked to quickly regain control of the landmine situation—to force it back into a controlled mode of slow violence—before the potential of the UXO threat became a reality.

Rhythmic temporalities of slow violence and emergency

To understand a crisis, we must consider how seemingly past crises re-emerge during a present emergency event and reshape the landscape. The power of the flood waters brought the quotidian, latent violence of the landmines to the center of flood response. There is no binary between an acute emergency event and the slow violences with which it interacts. Instead, the response to the interplay of water, land, and landmines reveals the different and, sometimes, inconsistent rhythms of threat and harm. The emergency of the floods—mixed with the threat of potential landmines in unknown locations—increased the urgency with which authorities and international aid responded to the need for evacuation. Finding and disarming landmines became another aspect of flood response. The importance of learning the mine locations increased with the demand for dealing with the flood event itself.

The speed of the water and the quickness with which its volume increased raised the level of concern about the landmines. The new unknown locations of the mines reified the urgency of their unknown temporality which is chronic until its acute explosion. What David Henig calls "the indeterminacy" of life shaped by landmines became more pronounced (2019). A mine may go off on its own due to deterioration (World Bank et al. 2014a) or it may go off when a person or animal steps on it. However, the potentiality that a mine will explode remains constant. As the flood shifted the temporality of the mines, the explosive potential of the mines shifted the temporalities of emergency preparedness and response to the floods.

In order for states, organizations, and individuals to safely continue toward flood recovery, authorities had to prioritize locating, marking UXO locations, and examining previously mine-free land for UXOs. While many of the mines moved during the flood are now marked, they, and others globally, lie in wait to be triggered. Fear of unknown harm by the UXOs also produces a slow violence for those who must pass through contaminated land. This "military waste" of UXO hinders the ability to restore livelihoods after the conflict, and thereafter any emergency event that reshapes the spaces of control that had allowed social and economic life to return after the combat event (Henig 2012). Further, this waste, in combination with natural hazards and geopolitical maneuvers by political actors, such as states, creates an additional layer of threat for current refugees passing through southeastern Europe entering the EU.

Conclusion: unpredictable threats along the Balkan route

Before the May 2014 floods, authorities expressed concern that refugees from countries were crossing through areas that had known mine contamination (Croatian Ministry of Interior 2014, in UNDP (HR RNA) 2015). At that time borders were not closed, and most refugees continued to travel from the south of Serbia to its border in the north with Hungary. While residents of BiH, Croatia, and Serbia were the most directly affected by the floods, refugees who traveled to the region from other countries were also affected. In 2014 most refugees were still passing through Serbia to go to Hungary. Some refugees in Serbia helped in the immediate rescue of people trapped in their homes (Interview E 2015). Yet, refugees were most affected by the floods due to the changes in the physical and political landscapes that occurred in the years after 2014. The floods reshaped the landscape of BiH, northern Serbia, and northeast Croatia, moving and hiding UXOs. Political decisions post-2014 made each of these regions a key part of the refugee path to the EU.

Serbia has remained an important transit country for refugees, especially since 2014. It was a key node in the path to the EU borders of Hungary and Croatia. When Hungary closed its border in 2016, refugees shifted direction to enter the EU through the Serbia-Croatia border instead (Minca et al. 2019; Umek et al. 2019). However, to avoid police violence they needed to travel away from the roads and so faced a new threat of triggering landmines as they traveled into Croatia (BBC 2015). As Croatia tightened its borders in 2017, refugees again shifted the direction of the unofficial Balkan route. Refugees entered BiH through Serbia. Like Serbia, BiH is in accession to join the EU, and at that time, its borders with the EU were less policed than those in Serbia. Due to the large number of UXOs in BiH, refugees face heightened risk from this particular kind of slow violence.[1]

These refugees have nothing to do with the conflicts that led to mine placement, and many were born long after the mines were in the ground. To discuss the multiple layers of violence on the bodies of refugees as they try to enter the EU (see Dhesi et al. 2018), we must also examine how natural hazard emergency events, such as floods, affect the type of slow violence wrought by UXOs that remain embedded in the landscape. This potential violence brought about by the placement of UXOs in the ground and the lack of funds to remove them adds an additional layer of threat to the refugees' journeys. While scholarship on the body and slow violence is growing (see Davies et al. 2017), future research will further engage with the intersection of slow violences and natural hazard events and how the interaction between them affect the vulnerable bodies of residents and refugees.

Slow violence alerts us to temporalities of harm and is a reminder that violence can occur at slow, often unseen, speeds. When any kind of slow violence and natural hazard event interact, there is an increased chance of that emergency event becoming a disaster. The interaction between landmines and the May 2014 floods in southeastern Europe demonstrates how the specific, uneven temporality of landmines is remade and dispersed through interaction with the flood. This chapter explores the temporalities of diverse crises as rhythms to acknowledge how they interact and intersect, reshaping the directions and effects of emergency events. Exploring how temporally different crises intersect deepens our understanding of the consequences that emerge and remain long after moments of emergency.

Acknowledgments

Many thanks to Kelly Chen, Sarah Friedman, Bob Kaiser, Ola Oladipo, and Jacob Remes for thoughtful feedback on this chapter. Research for this chapter was supported by the Scott Kloeck-Jensen Pre-dissertation Travel Grant at the University of Wisconsin-Madison and by the Title VIII Combined Research and Language Training Program, which is funded by the U.S. State Department, Title VIII Program for Research and Training on Eastern Europe and Eurasia (Independent States of the Former Soviet Union) and administered by American Councils for International Education. The opinions, findings, and conclusions stated herein are the author's own and do not necessarily reflect those of either the U.S. Department of State or American Councils. This chapter was written with support provided by the Graduate School at the University of Wisconsin-Madison, part of the Office of the Vice Chancellor for Research and Graduate Education, with funding from the Mellon Foundation.

Note

1. In March 2021, one asylum seeker was killed in Croatia after stepping on a land-mine and four others were injured (Tondo 2021).

References

Adams, Vincanne (2013), *Markets of sorrow, labors of faith: New Orleans in the wake of Katrina*. Durham, NC: Duke University Press.

Anderson, B., K. Grove, L. Rickards, and M. Kearnes (2019), 'Slow emergencies: temporality and the racialized biopolitics of emergency governance', *Progress in Human Geography*, n.p., https://doi.org/10.1177/0309132519849263

Anderson, K. (2000), 'The Ottawa convention banning landmines, the role of international non-governmental organizations and the idea of international civil society', *European Journal of International Law*, 11(1), 91–120.

Baker, Catherine (2018), *Race and the Yugoslav region: postsocialist, post-conflict, post-colonial?* Manchester: Manchester University Press.

Bakker, K. (2005), 'Katrina: the public transcript of "disaster"', *Environment and Planning D: International Journal of Urban and Regional Research*, 23(6), 795–801.

Bankoff, G. (2007), 'Comparing vulnerabilities: toward charting an historical trajectory of disasters', *Historical Social Research/Historische Sozialforschung*, 103–114.

Bolton, Matthew (2010), *Foreign aid and landmine clearance: governance, politics and security in Afghanistan, Bosnia and Sudan*. London: IB Tauris.

British Broadcasting Corporation (BBC) (16 September 2015), 'Migrant crisis: Croatia mines warning after border crossing', accessed 20 June 2019 at https://www.bbc.com/news/world-europe-34268043

Cannon, T. (1994), 'Vulnerability analysis and the explanation of "natural" disasters', *Disasters, Development and Environment*, 1, 13–30.

Chick, K. (2014, 5 July), 'Bosnia's flood clean-up brings a hazardous wrinkle: land mines', *Christian Science Monitor*, accessed 10 October 2019 at http://www.csmonitor.com/World/Europe/2014/0705/Bosnia-s-flood-clean-up-brings-a-hazardous-wrinkle-land-mines

Christian, J.M. and L. Dowler (2019), 'Slow and fast violence: a feminist critique of binaries', *ACME: An International E-Journal for Critical Geographies*, 18(5).

Clayfield, M. (2017, 14 October), 'Bosnia may never be clear of landmines', Australian Broadcasting Corporation, accessed 1 July 2019 at https://www.abc.net.au/news/2017-10-15/bosnia-may-never-be-clear-of-land-mines/9029692

Crowther, G. (2008), 'Counting the cost: the economic impact of cluster munition contamination in Lebanon', report for *Global CWD Repository*, 1100.

Davies, T. (2018), 'Toxic space and time: slow violence, necropolitics, and petrochemical pollution', *Annals of the American Association of Geographers*, 108(6), 1537–1553.

Davies, T., A. Isakjee, and S. Dhesi (2017), 'Violent inaction: the necropolitical experience of refugees in Europe', *Antipode*, 49(5), 1263–1284.

Dhesi, S., A. Isakjee, and T. Davies (2018), 'Public health in the Calais refugee camp: environment, health and exclusion', *Critical Public Health*, 28(2), 140–152.

Egel, N. (2020), 'The Trump administration approved the U.S. use of land mines. That's a step back for global campaigns to ban their deployment', *The Washington Post*, accessed 25 February 2020 at https://www.washingtonpost.com/politics/2020/02/11/trump-administration-okd-us-use-landmines-thats-step-back-global-campaigns-ban-their-use/

Field, J. and I. Kelman (2018), 'The impact on disaster governance of the intersection of environmental hazards, border conflict and disaster responses in Ladakh, India', *International Journal of Disaster Risk Reduction*, **31**, 650–658.

Henig, D. (2012), 'Iron in the soil: living with military waste in Bosnia-Herzegovina', *Anthropology Today*, **28**(1), 21–23.

Henig, D. (2019), 'Living on the frontline: indeterminacy, value, and military waste in postwar Bosnia-Herzegovina', *Anthropological Quarterly*, **92**(1), 85–110.

Hyndman, Jennifer (2011), *Dual disasters: humanitarian aid after the 2004 tsunami*. Sterling, VA: Kumarian Press.

International Campaign to Ban Landmines (ICBL) (2019), 'The Ottawa Treaty: campaign to ban anti-personnel mines', accessed 30 October 2019 at http://www.icbl.org/en-gb/the-treaty/treaty-in-detail/treaty-text.aspx

Kakissis, J. (2014, 20 May), 'Balkan floods expose deadly mines from 1990s civil war', *National Public Radio*, accessed 1 August 2019 at https://www.npr.org/sections/thetwo-way/2014/05/20/314379214/balkan-floods-expose-deadly-mines-from-1990s-civil-war

Katz, C. (2008), 'Bad elements: Katrina and the scoured landscape of social reproduction', *Gender, Place & Culture*, **15**(1), 15–29.

Kelman, I. (2007), 'Hurricane Katrina disaster diplomacy', *Disasters*, **31**(3), 288–309.

Kim, E.J. (2016), 'Toward an anthropology of landmines: rogue infrastructure and military waste in the Korean DMZ', *Cultural Anthropology*, **31**(2), 162–187.

Landmine and Cluster Munitions Monitor (2012), 'Serbia', *Landmine and Cluster Munitions Monitor*, accessed 1 June 2019 at http://archives.the-monitor.org/index.php/cp/display/region_profiles/theme/2168

Landmine Monitor (2018), 'Casualties', *Landmine and Cluster Munitions Monitor*, accessed 1 June 2019 at http://www.the-monitor.org/en-gb/reports/2018/landmine-monitor-2018/casualties.aspx

Matic, J. and R. Smajilhodzic (2014), 'Landmine explodes in Bosnia dangerous cleanup begins', *Sydney Morning Herald*, accessed 1 August 2019 at https://www.smh.com.au/world/landmine-explodes-in-bosnia-as-dangerous-clearup-begins-after-massive-floods-20140522-zrkhl.html

Minca, C., D. Šantić, and D. Umek (2019), 'Managing the "refugee crisis" along the Balkan route: field notes from Serbia', in C. Menjivar, M. Ruiz, and I. Ness (eds.), *The Oxford Handbook of Migration Crises*. Oxford: Oxford University Press, pp. 50–65.

Monin, Lydia and Andrew Gallimore (2011), *The Devil's Gardens: the story of landmines*. London: Random House.

NASA (2014), 'Severe flooding in the Balkans', *NASA Earth Observatory*, accessed 5 August at https://earthobservatory.nasa.gov/images/83697/severe-flooding-in-the-balkans

Nixon, R. (2007), 'Of land mines and cluster bombs', *Cultural Critique*, **67**(1), 160–174.

Nixon, Rob (2011), *Slow violence and the environmentalism of the poor*. Cambridge, MA: Harvard University Press.

Nobert, S., Rebotier, J., Vallette, C., Bouisset, C. and Clarimont, S. (2017), 'Resilience for the Anthropocene? Shedding light on the forgotten temporalities shaping post-crisis management in the French Sud Ouest', *Resilience*, **5**(3), 145–160.

Norwegian Peoples Aid (2008), *Yellow killers: The impact of cluster munitions in Serbia and Montenegro*.

Norwegian Peoples Aid (2014, 20 May), 'NPA and the flood in Bosnia-Herzegovina', accessed 1 August 2019 at https://www.npaid.org/News/News-archive/2014/NPA-and-the-flood-in-Bosnia-Herzegovina

O'Keefe, P., K. Westgate, and B. Wisner (1976), 'Taking the naturalness out of natural disasters', *Nature*, **260**, 566–567.

O'Lear, S. (2016), 'Climate science and slow violence: a view from political geography and STS on mobilizing technoscientific ontologies of climate change', *Political Geography*, **52**, 4–13.

Pelling, M. and K. Dill (2008), 'Disaster politics: from social control to human security. Environment', *Development and Politics Working Paper Series*, 1–25.

Pulido, L. (2000), 'Rethinking environmental racism: white privilege and urban development in Southern California', *Annals of the Association of American Geographers*, **90**(1), 12–40.

Schwartzstein, P. (2014, 21 December), 'For leopards in Iran and Iraq, land mines are a surprising refuge', *National Geographic*, accessed 9 October 2019 at https://www.nationalgeographic.com/news/2014/12/141219-persian-leopard-iran-iraq-land-mine/

Serbia Mine Action Center (SMAC) (2014, 23 June), 'Update on Article 5 implementation'. Letter to the Anti-personnel Mine Ban Convention conference in Mozambique, accessed 20 August 2019 at https://www.maputoreviewconference.org/fileadmin/APMBC-RC3/tuesday/07c_CLEARING_MINED_AREAS_-_Serbia.pdf

Serbia Mine Action Center (SMAC) (2019, November), 'Mine situation', accessed 1 November 2019 http://www.czrs.gov.rs/eng/minska-situacija.php

Smallwood, Peter D. and Todd R. Lookingbill (2019), 'Battlefields and borderlands: the past, present and future of collateral values', in Peter D. Smallwood and Todd R. Lookingbill (eds.), *Collateral values: the natural capital created by landscapes of war*. Cham, Switzerland: Springer, pp. 263–270.

Smith, N. (2006, 11 June), 'There's no such thing as a natural disaster'. Understanding Katrina: perspectives from the social sciences, accessed 1 August 2019 at http://blogs.ubc.ca/naturalhazards/files/2016/03/Smith-There%E2%80%99s-No-Such-Thing-as-a-Natural-Disaster.pdf

Stec, S., J. Kovandžić, M. Filipović, and A. Čolakhodžić (2011), 'A river ran through it: post-conflict peacebuilding on the Sava River in former Yugoslavia', *Water International*, **36**(2), 186–196.

Tondo, L. (2021), 'Croatia: landmine from 1990s Balkan wars kills asylum seeker', *The Guardian*, accessed 20 March at https://www.theguardian.com/world/2021/mar/07/croatia-landmine-from-1990s-balkans-war-kills-asylum-seeker

Tyner, James A. (2010), *Military legacies: a world made by war*. New York: Routledge.

Umek, Dragan, Claudio Minca, and Danica Šantić (2019), 'The refugee camp as geopolitics: the case of Preševo (Serbia)', in Maria Paradiso (ed.), *Mediterranean mobilities: Europe's changing relationships*. Cham, Switzerland: Springer International Publishing, pp. 37–53.

United Nations Development Programme (UNDP) (HR RNA) (2015), *UNDP mine action recovery needs assessment for flooded areas in eastern Croatia*. UNDP.

United Nations Disaster Assessment and Coordination Team (UNDAC) (2014), 'End of Mission Report. Mission to Serbia – Floods 18–31 May 2014', accessed 1 October 2019 at http://www.undp.org.rs/download/Final%20UNDAC%20Report%20-%20Serbia%20Floods%20May2014.pdf

Unruh, J.D., N.C. Heynen, and P. Hossler (2003), 'The political ecology of recovery from armed conflict: the case of landmines in Mozambique', *Political Geography*, **22**(8), 841–861.
Wisner, B. and H.R. Luce (1993), 'Disaster vulnerability: scale, power and daily life', *GeoJournal*, **30**(2), 127–140.
World Bank, United Nations Development Programme, the European Union and BiH government (2014a), *Bosnia and Herzegovina (BiH) Recovery Needs Assessment (RNA)*, accessed 1 October 2019 at https://www.gfdrr.org/sites/default/files/BiH-rna-report.pdf
World Bank, United Nations Development Programme, the European Union and Serbian government (2014b), *Serbia Floods Reported Needs Assessment* (SR RNA), accessed 1 June 2019 at http://www.sepa.gov.rs/download/SerbiaRNAreport_2014.pdf
World Health Organization (WHO) (2014, June 3), 'Floods in the Balkans: Bosnia and Herzegovina, Croatia and Serbia: situation report no. 2', *Situation Report*, **2**, 2–3.

4

Complicating the role of sight: photographic methods and visibility in slow violence research

John Paul Henry

Introduction

This chapter connects the experiences of those living near hazardous industrial sprawl to the problematic relationship of sight in understanding industrial toxicity. Industrial practices like flaring, or the burning of excess chemicals, and other dramatic events divert attention away from ongoing toxic emissions and the legacy genome-altering chemicals. The inability of invisible toxins to garner sustained attention is central to Rob Nixon's theorization of slow violence (Nixon 2011). Rather than relying on quantifiable metrics such as emission data and cancer rates, this chapter instead seeks to understand slow violence from the perspective of everyday life.

The challenge, as Nixon suggests, is how to make visible the taken-for-granted toxic exposure and industrial practices that have become part of the everyday fabric of life along the edges of hazardous industrial facilities, or fenceline communities. To better understand experiences of slow violence, this chapter connects a weave of stories, personal trajectories, and industrial practices, which converge in Calvert City, Kentucky. Each situated perspective provides a different method for viewing, understanding, and articulating experiences of slow violence in the context of industrial pollution. Considering the normalization of these industrial practices and the invisible yet toxic nature of historic and ongoing toxic releases in Calvert City, this chapter seeks to answer the following questions: How can experiences of slow violence be articulated through creative collaboration? How does the role of sight affect one's ability to understand slow violence? How can understanding everyday experiences of those living in toxic places better inform our understanding of slow violence?

Everyday experiences of toxic exposure are brought forth in this study through a collaborative creative process. The chapter offers a detailed construction of a critical, creative methodology (Rose 2016) with techniques developed from the *photowalk* concept (Cannuscio et al. 2009), and adapted for rural geographies, called the *photodrive* (Henry 2020). Through photowalks, researchers and collaborators co-navigate everyday spaces and share the creative process of making photographs. This method can be adapted to rural geographies through the use of a car in the form of a photodrive. In both instances, the researcher's goal is to focus on aspects of everyday life illuminated through the co-navigation of space.

Methodologies that center everyday experiences help uncover the gendered, classed, and often racialized differences in toxic exposure. This grounded method makes visible both the daily life of fenceline communities and embodied forms of knowledge related to slow violence. This chapter provides a roadmap for using creative methods as a mode of intervention into toxic places. The chapter concludes by problematizing the reliance of sight in understanding invisible toxins and other overlooked yet violent practices.

The call to make slow violence visible

To understand slow violence, it is necessary to understand how a wide range of practices is construed as violent. Practices that routinely limit or deny others the ability to reach optimum health, either nutritionally, biologically, or emotionally, are acts of violence (Galtung 1969, p. 168). Galtung particularly foregrounded the violent effects of different state apparatuses built into systems of governance, or structural violence. Structural violence over time becomes viewed as natural in the structure of society. Uneven access to healthcare in the United States is a form of structural violence that allows certain individuals access to wellbeing while threatening others with debilitating medical debt. Other forms of violence, however, are less traceable but are evident in practices that foster or perpetuate unevenly harmful outcomes: state-level legislation, corporate pollution, and familial emotional abuse. This list is by no means exhaustive, yet it illuminates how both institutional and personal networks may cause and obscure violent practices.

A focus on slow violence further develops our understanding of violence by recognizing the varied, often hidden, and particularly long temporalities through which practices unfold. As Galtung sought to foreground networks of power, Nixon foregrounds time. Importantly, especially in relation to toxic exposure, slow violence makes sustaining life increasingly difficult (Nixon 2011, p. 3). Therefore, incremental exposure to toxins gradually reduces quality of life and creates geographies of slow violence as communities are often unknowingly exposed to industrial chemicals.

The 20th century is replete with examples of gradually degrading the ability to sustain life. While promising to improve economic conditions, chemical facilities and the political-regulatory regimes have time and again chosen to situate toxic industries at the nexus of classed and racial geographies (Bullard 2001). Predominately African American communities in the U.S. South are much more likely to be situated adjacent to toxic incinerators than White, more affluent communities. These geographies of slow violence are not isolated, unrelated practices.

Geographies of slow violence are tied to economic and political networks of decision-making and control. For instance, at a southern Illinois uranium enrichment facility the effluent discharge of uranium is self-reported to the state Environmental Protection Agency, which allows discharge into the Ohio River (Public Notice 2016). On the southern Mississippi River petrochemical corporations draft public notice statements to be presented as coming from the public sector (Davies 2018, p. 9). These two examples provide context into the political networks of power that decide appropriate practices through which populations are exposed to toxicity. However, assigning blame to practices of slow violence, for instance the incremental, daily release of toxins over decades, is difficult in juridical terms (Romero et al. 2017). While it is often impossible to prove intentionality by linking toxic exposure to specific actors, the practice of land zoning and even specific industrial practices *can and should* be linked to actors. Such is the case in this chapter concerning industrial practices in Calvert City which date back to the 1950s. A more in-depth discussion of industrial practices will be presented later in this chapter.

While it is necessary to identify networks of power in order to hold corporations and politicians accountable, this chapter's focus is to answer Nixon's call to make slow violence visible. Geographic inquiry is well-suited to understanding slow violence as attention to these networks inherently spatializes violent practices. Like chemicals leeching into groundwater, the effects of slow violence are not immediately felt or known. Listening to people and trying to see what they see can give insights not only into the networks of power but also into everyday lived experiences.

Toxic geographies continue to diminish life across the planet. Deregulation and violent practices threaten to exacerbate or make new invisible crises (Popovich et al. 2020). Slow violence requires sustained focus if researchers are to make meaningful interventions. New methods of visualizing and engaging with slow violence are needed to overcome the abbreviated attention spans associated with social media. The following section outlines the role of photography, and other creative practices, in making slow violence visible.

The creative process in slow violence research

Making slow violence visible through the use of photography might seem like an oversimplified solution, yet this section outlines the value of the *creative process*, rather than the end result, both for collaborators and researchers. One of Nixon's central themes concerning slow violence is a challenge of representation and visibility. As slow violence occurs "gradually and out of sight" it is even more pressing to develop research methods that incite a sense of urgency (Nixon 2011, p. 2). Photographic methods create representations of everyday places as captured and seen from the perspective on the bodily experience. While this practice is valuable in its own right, and *may* contribute to creating a sense of urgency, this chapter demonstrates that the space created through a shared creative process helped me as a researcher to recognize and see overlooked and normalized everyday experiences. Through sharing these normalized and overlooked everyday experiences we can begin to change narratives of toxic places.

At the core of the difficulty of representing slow violence is that it is often inherently obscured by time. After all, as Nixon reminds us of the early 20th-century philosopher and ecologist Aldo Leopold's ethical proposition, "we can be ethical only toward what we can see" (Nixon 2011, p. 15). This stance has certainly become problematic as knowledge of radiation and bioaccumulation highlights how the invisible can be just as deadly as the visible. Yet, in the eyes of the public at large, that which is seen on the phone, computer, or television screen is too often equated with truth. This makes answering Leopold's following question even more relevant in today's world. "What then, in the fullest sense of the phrase, is the place of seeing in the world that we now inhabit?" (Nixon 2011, p. 15). Seen from another perspective, if slow violence is rendered in often invisible or in hidden ways, what is the role of sight? Here,

my aim is to address this question in terms of the knowledge-creating potential, not through the final creation of a photographic image, but in the situated
and creative practice of *seeing together*. Given that slow violence is invisible,
photography created of everyday life may not spark outrage or catalyze radical
movements. What this method does, however, is create forms of knowledge
that are otherwise unspoken, or dismissed. It allowed me to see the subtle ways
in which slow violence manifests. Geographers have used similar methods to
better understand and represent our world.

The last two decades of creative geographic practice infused geographic
thought with interdisciplinary practices of the humanities and the visual
arts. Creative geographic research strategies often combine geographers and
artists, including poets, musicians, filmmakers, photographers, sculptures,
and conceptual artists (de Leeuw and Hawkins 2017; Garrett 2010; Kindon
2003). Geographers can practice creativity as well, taking up canvas and paint
to better understand ways of knowing, problematize representation, and as
methods of intervention (Hawkins 2015). The resulting creative practices have
helped shift the epistemological balance from quantifiable 'truths' as accepted
knowledge to a centering of more relative, lived experiences as likewise valid
knowledge. This shift occurred largely because creative methods eschew the
rigid and often *a priori* categorizations established by agents of power and
instead value an open-ended approach to knowledge creation.

A critical creative practice must be concerned with how the creative process
remakes human relationships with each other and space. Creative works
"make and remake spaces, places, and human relationships" (de Leeuw and
Hawkins 2017). The practices of creating, reflecting on, and viewing creative works affect the cultural construction of space (Massey 2005). Creative
works, then, should be seen as information created in historical contexts as
a co-constructive force, informed by society but also shaping it (Pickett et al.
2019). Geographers should be attuned to the cultural traces that lead collaborators, and the researchers themselves, to view the world in certain ways.

It is necessary to bridge the broad category of critical creative practices,
spanning from poetry to dressmaking, to the more specific visual practices,
including photography. A subset of the creative methods, photography can
be a *visual research method* which addresses research questions and creates
knowledge through the creation of new images. Rose identifies three concepts
necessary for the construction of a critical visual methodology (Rose 2016,
p. 22). A critical approach must first take images seriously; it must, second,
consider the effects of images and their distribution; and, third, it must be con-
siderate of the researcher's way of looking (Rose 2016, p. 22). The last criteria,

also thought of as reflexivity, is integral in understanding how the researcher's presence influences the interview and the creation process. Researchers are likely to interpret photos quite differently than participants. Unique personal experiences, cultural perspectives, and genealogies all influence how we view place. An attentiveness to these divergent perspectives, and the influence of the researcher on the interview setting, are important aspects to consider when employing visual research methods.

The goal of sharing the creative process should be knowledge creation and a convergence of perspectives. Employing creative methodologies without a critical stance risks the "instrumentalization of knowledge" (de Leeuw and Hawkins 2017). An uncritical approach, likewise, risks positioning photographs as "isolated slices of reality" rather than a medium which bridges experiences of the researcher and collaborator (Davies 2013, p. 134). The nuance here is important and points to the value of the shared creative exploration of space. The end result (i.e. photograph, video, participatory map, etc.) is always bound in power imbalances and subject to audiencing beyond the researcher's control. Yet the urgency of countless geographies of slow violence enacted across the planet necessitates more diverse and creative methods for garnering understanding, attention, and ultimately action.

Photographic visual research methods

The following methods are useful for responding to and making visible geographies of slow violence: photo elicitation, participatory photography, and photowalks. All three methods are based on the photographic image, either provided or created. The three previously mentioned methods revolve around the idea that the photographic image provides a means of sharing common experience through sight. Yet, photowalks are especially suited for bringing forth hidden everyday experiences of slow violence. One's ability to create aesthetically pleasing pictures is not necessary for these methods as the value also lies in the critical reflection of the content and the critical reflection of the creation process itself.

Photo elicitation is the oldest and most time-efficient method which uses photographs as part of the interview process. Simply, photographs are used as prompts during interviews and can function as aids providing open-ended, interpretive value. Elicitation photos may come from found images, supplied by the participant, or may be participant or collaboratively created between the collaborator and researcher (Rose 2016, p. 304). Photo elicitation evokes richer, and at times encyclopedic knowledge, including description, explanation, analysis, emotion, and affect (Rose 2016). Interviews of Japanese-Canadians regarding World War II internment camps became "more intense and emotional as photos stirred deep and often painful memories" (Rose 2016, p. 305). It is from elicitation interviews that researchers should develop codes for data, patterns, ways of knowing, and embodiments revealed through the interview process.

While photo elicitation interviews are interpretive, participatory visual methods allow creative agency on the part of the participant. The participant dictates what to photograph, how to photograph, creates the images themselves, and through interviews gives meaning to the images created. Participants have used photography to share the realities of life near the Chernobyl Exclusion Zone in which the toxicity was regulated by the state yet remained invisible to everyday Ukrainians (Davies 2013, p. 117). A participant photograph of a bowl of half-prepared fish sparked discussion of illegal fishing in the Exclusion Zone in which likely contaminated fish is obtained in exchange for potatoes. "'I have no money to pay them, only potatoes. If I have the money to pay them then I do'" (Davies 2013, p. 133). In this instance, the participant photograph and subsequent interview illustrated everyday practices that develop into pathways for toxic exposure.

A third visual method, the photowalk, blurs the lines between researcher and collaborator, simultaneously complicating the creative process yet bringing disparate world perspectives into conversation in place. Photowalks are walking interviews that allow the photographic process to be shared between researcher and collaborator. Simply, photowalks allow researcher and collaborator to create pictures together. Photowalks provide more than simple means to representation as the photowalks promote a unique blending of perspectives as researcher and collaborator co-navigate spaces, mimicking a dialectic exchange of knowledge and experiences. For instance, during a photowalk interview in Philadelphia, researchers identified a community garden for its perceived benefits while collaborators related years of community resistance against drug dealers that once occupied public housing formerly located on the garden plot (Cannuscio et al. 2009, p. 557). An important distinction is that photowalks are inherently a collaborative process, rather than participatory, as open-ended conversations and observations give rise to surprises. Both perspectives are challenged as divergent senses of place are shared in the mutual exploration of a situated becoming of place (Pred 1984, p. 292). In other words, the place of the interview matters. Photowalks and subsequent photo elicitation interviews acknowledge and value the everyday practices and knowledge of collaborators as legitimate forms of knowledge.

The previous three visual methods may be used to better understand divergent perspectives of place, which is crucial to understanding everyday experiences of slow violence. Photo elicitation, participatory photography, and photowalks form a set of methodological tools that, when used in tandem, may help researchers go beyond quantification reports to understand lived experience in toxic places. The following section describes the chemicalization of western Kentucky. The legacy of toxic emissions and dumping, coupled with ongoing fugitive emissions and regular chemical flaring, creates a need for further understanding everyday experiences in this rural area developed around the influx of chemical facilities.

Case study using photographic methods

A series of chemical facilities in the Lower Tennessee River flood plains of western Kentucky fit broader patterns of toxic waste dumping in the American South in which class influences the likelihood of toxic exposure. While Robert Bullard (2001) demonstrated the racialized spatiality of toxic waste dumping and incineration throughout the South, in western Kentucky the geographies of industrial slow violence are more closely reflected in Lerner's concept of a sacrifice zone (Lerner 2010). Sacrifice zones connect patterns of industrial development and subsequent toxic releases to low socio-economic residential areas.

Economic development in the Tennessee River valley of western Kentucky remained negligible until the 1940s when the Tennessee Valley Authority began work to dam the Tennessee River in order to create electricity. Chemical manufacturers flocked to the flood plains north of Calvert City, Kentucky, to take advantage of the 'cheap' power. This concentrated area of industrialization is known as the Calvert City Industrial Complex, the focal point for my research.

A pattern of toxic releases followed the development of the Calvert City Industrial Complex. In the 1980s, a local reporter broke the story that chemical waste had been buried in unlined pits in the flood plain from the 1950s through the 1980s (McDonald 2017). Furthermore, the now-shuttered incineration company, Liquid Waste Disposal, Inc. (LWD), handled and incinerated toxic waste for decades (Enforcement 2016). LWD discontinued operations in 2003. The former LWD property is now a Superfund Site as high levels of the following toxic substances were found on the property: "volatile organic compounds (VOCs), such as TCE, benzene, toluene, vinyl chloride, and xylenes, as well as toxic metals, such as arsenic, lead and chromium. High levels of dioxins were also identified in the incinerator ash abandoned at the site" (Enforcement 2016). Dioxins, which were emitted by the LWD incinerator for decades, bio-accumulate in the fatty tissues of animals, including fish, but do not kill them (Schwartz et al. 1983; White and Birnbaum 2009). "Dioxins have proven to be developmentally toxic, immunotoxic, neurotoxic, and hepatoxic" and, impor-tantly, may cause impairments in immune function without producing overt signs of toxicity (Schwartz et al. 1983, p. 49). As recently as 2010 (Sturgis 2012), Calvert City was identified as one of the worst dioxin polluters in the U.S. for surface water disposal of dioxins.

The eventual shuttering of LWD reduced *visible* emissions from the Calvert City Industrial Complex. More recent industry focus and federal resources have turned towards tracking the underground chemical plume spreading beneath the Tennessee River. EPA data (Release Reports 2019) suggest Calvert City Industrial Complex practices are far from clean, as fugitive emissions, emissions *not* filtered nor flared from point sources, rival emissions in Cancer Alley in southern Louisiana.

Given the lack of sustained attention given to the bioaccumulative nature of chemicals handled at the Calvert City Industrial Complex *and* the knowledge of fugitive emissions of known carcinogens, these circumstances suggest the need for continued scrutiny for the wellbeing of local residents. In this light, this study considers the history of pollution and the ongoing industrial prac-tices at the Calvert City Industrial Complex and seeks to address the following questions: (1) How can experiences of slow violence be articulated through creative collaboration? (2) How does the role of sight affect one's ability to understand slow violence? (3) How can understanding everyday experiences of those living in toxic places better inform our understanding of slow violence? These questions seek to create knowledge that values lived, everyday experi-ence that may otherwise be deemed not worthy by local media and *a priori* studies.

The following comments are drawn from my 2019 fieldwork utilizing open-ended interviews which suggest a conflicted understanding of industrial flaring, or burning off excess chemical emissions. Stories about Calvert City in the absence of the notorious LWD incinerator emissions are dominated by flaring narratives and concerns. The visibility of chemical flaring redirects community attention and narratives from the bioaccumulative property of dioxins handled on the complex and ongoing fugitive emissions to the practice of flaring. Flaring is viewed as problematic but an improvement to the health and safety of the community. Marshall County and neighboring Livingston

County residents report various perspectives of benzene flaring, a common hydrocarbon and known human carcinogen (CDC 2018). "For my farm, when they're burning off benzine at night, which is against regulations, you can definitely see it because the whole sky glows off in the east. It's usually red and sounds like thunder," a collaborator and long-time resident told me. He called the local law enforcement to complain. "So what? They're just burning benzene," he was told.

Another local resident said when she was a newcomer to Calvert City the flares woke her from bed. She went out to her front porch for a look. "You could see that flame in the sky and hear the roar of the jet engine. It sounds like a jet engine when that goes off. It's kind of terrifying when you look up into the sky and there's a big black cloud of smoke." Her mother, a long-time resident, reassured her, "Oh, they're just burning chemicals off" (Henry 2020, p. 237). This routinization of the chemical flaring has helped to create a narrative surrounding the practice of flaring. Since flaring is generally believed to be safe, or at least better than the unfettered release of chemicals, most collaborators viewed flaring as an improvement. Living in close proximity to chemical facilities has become routinized as many people live their entire lives in Marshall County. Critical visual methodologies provide a way of breaking through the acceptance by centering everyday experiences so that we may better understand the relative and unique experiences of slow violence. The previous anecdotes provide insight into the role of sight in relation to understanding slow violence.

During my initial visit to the Calvert City Industrial Complex my gaze, too, turned towards the flaring exhaust stacks, through which excess chemicals are funneled and burned. It became clear that my understanding of toxicity, and its sources, was much different than the nuanced, everyday knowledge of my collaborators. Through this work, I came to recognize the difference in stories portrayed through traditional interview settings compared to when collaborators and I visited places together. The following narratives are of a retired pipefitter and his spouse who share a combined total of more than 90 years of experience living near the Calvert City Industrial Complex.

The retired pipefitter, his spouse, and I sat in the Calvert City Dairy Queen, our conversation occasionally punctuated with brief visits by their friends. The pipefitter settled into telling stories of life working in chemical facilities. One of his friends was burned in a chemical fire; another was crushed in an accident. Eventually the conversation turned to the pipefitter as he pulled up his sleeve to show me a mark of discoloration on his arm. "Is it cancerous?" I ask. He replies, "Not yet, but it turns into cancer. That was over at Goodrich ..."

(Henry 2020, p. 235). He had worked in these Calvert City plants since 1973. Over the years he built such a familiarity with the chemical facilities that he was able to use the table to sketch a mental map. The well had been plugged, but he said EPA officials had not bled the pressure off. That is when the contaminated groundwater sprayed him. "What happened when you reported it?" I asked. "They said, 'Shut your face or you won't have a job.' And work was real slender back then. I don't think I was making more than 16, 17 dollars an hour. Wasn't worth it, looking back at it." In other words, the pipefitter said he was silenced from sharing this story by threat of financial retribution. While my interview with the pipefitter highlights skewed power relations between contract workers and industrial chemical facilities, the conversation's focus revolved around professional life. Substantially more, and at times surprising, knowledge of everyday life was brought forth when these collaborators agreed to show me around the Calvert City Industrial Complex to create photographs.

As our vehicle pulled away from the restaurant heading toward the industrial complex, the tone of the focus of the conversation became more casual. Within minutes the pipefitter's spouse mentioned concerns of drinking the municipal water. Oily rings appear in the toilet, sink, and even the dog's water dish, she said, only a few days after being cleaned. That's why they only drink bottled water.

Circling through a trailer park across the street from the industrial complex something jogged the pipefitter's spouse's memory. "Another thing about living in Calvert, everything corrodes really fast. Like the outside brass on our house. And we live five miles from the plants," she said. Just then we pulled up to a stop sign. To the north, facing the chemical facility, rust peeled away from the stop sign. To the south, facing away from the facilities, the metal remained smooth, untouched. "This is what he's talking about living on this side of the tracks. No one wants to live on this side of the tracks in Calvert City," the pipefitter's spouse said. The pipefitter agreed. "But if you was Black I definitely wouldn't come down here," he said. "Because I can take you to where they got the flag usually, and it's a confederate flag. It's Ku Klux Klan headquarters of western Kentucky. Old Gilbertsville" (Henry 2020, p. 237). This surprising revelation was reinforced by a 2015 recruitment campaign by the KKK that left fliers and candy throughout multiple towns in Marshall County, where the industrial complex is situated (Fritz 2015). More than 98% of Marshall County residents identify as White while only 0.2% identify as Black. The surrounding six counties average nearly 6% who identify as Black. Industrial facilities in western Kentucky are a primary source of economic mobility, especially in the case of Calvert City. This anecdote about KKK activities speaks to underlying social structures impeding everyday access to economic mobility for Black citizens in this region.

Because of time constraints, not all collaborators were able to conduct photowalks or photodrives. For those collaborators with time constraints, I asked how to photograph toxic places, particularly in the most mundane or ordinary of places. Importantly, these questions of *what to photograph* prompted, at times, far more detailed responses than interview questions regarding their personal experiences. Based on collaborators' suggestions, I then visited the suggested places and created photographs. This process allowed me to better understand, and document, the otherwise hidden places of everyday life adjacent to the chemical complex. These photos were later used in elicitation interviews in which participants offered their own meanings or interpretations of those photographs.

The emphasis here is on process and generating a creative exercise in which participants are asked to think about everyday places. Most collaborators were initially interested in representations of industrial infrastructure, aerial photography for scale, and landfills. My questions for this elective phase of the research project were instead focused on overlooked, everyday experiences. What do you see while driving to work? What sort of industry can you see or smell from your house? When employing this method, researchers should be attuned to why certain places are suggested. How did the collaborators interact with these places in everyday life? It is in these overlooked, everyday stories that nuanced understandings of slow violence can be made visible.

For instance, a long-time resident of Marshall County, who for a short time worked at the Calvert City Industrial Complex, did not view pollution from the chemical industries as problematic. When asked how to photograph pollution associated with the chemical facilities, he turned his attention elsewhere. A part-time commercial fisherman for decades, he recommended not eating fish out of specific areas of the local lakes and river systems, the dam locks, or marinas associated with the barge industry. "Where the boats are sitting, they leak oil and they leak this, and they leak that. Anywhere you have dead water instead of moving water. If there is a well or something leaking, then the fish would probably have a better chance of having chemicals in them, but over all you never see dead fish floating. If the water was toxic you would see fish floating," he said. The interviewee suggests that toxicity of any concern would lead to the death of fish. Research suggests otherwise. "Sport-caught fish is the major source of dioxins for the general population" (Schwartz et al. 1983, p. 50). His perception of toxicity was linked to the *visibility* of dead fish or, likewise, obvious signs of pollution. This example brings into light the complicated relationship between sight and toxicity. Even for many folks living near industrial facilities their entire life, the issue of hidden toxins is unclear. To better understand slow violence, it is important to develop knowledge about how practices of slow violence are misunderstood or dismissed because of a primary reliance on visibility to determine safety.

Conclusion: complicating a reliance on visuality

Geographers and their work could benefit from the reflexive practice of interrogating place through creative collaboration. While messy, the resulting knowledge exchange is markedly different than that of a typical interview setting. In this study, collaborative image creation and co-navigating toxic places helped make visible societal structures and everyday experiences that otherwise might be overlooked. For example, while conducting a photodrive, stories of local KKK networks were foregrounded, raising questions of equitable and safe access to employment. Similarly, while co-navigating toxic spaces, visual cues prompted a collaborator to described everyday observations of corrosive air in her fenceline community. Visual cues and visibility alone are not enough to fully understand geographies of slow violence.

In centering the role of sight in understanding slow violence this study demonstrates how a reliance on visibility alone can lead to an uncritical acceptance of toxic realities. Slow violence is often invisible. In the case of Calvert City, decades of improper chemical incineration and continued emission of dioxins have created a geography of slow violence which induces change at the genetic level, yet is unseen to the naked eye. As a fisherman and former chemical plant employee shared, he does not believe that local fish could be toxic because they do not exhibit visible symptoms. This overreliance on sight as a means of understanding risk has led to the uncritical acceptance of a history of toxic emissions and ongoing fugitive emissions.

Yet, what is the place of photography, seeing and recording through a camera, in understanding slow violence? Resulting photographs in fenceline communities like Calvert City often result in mundane depictions of everyday life. These photographs are not likely to spark outrage. However, when asking collaborators how to photograph mundane places that might be considered toxic, the question prompted reflection beyond the obvious. Answers were varied and showed that a collaborative creative process can provide more insight than the image itself. Indeed, the results of this project support the claim that creative methodologies help bring forth *different* forms of knowledge.

The industrial practice of flaring, which becomes normalized over time, can affect how communities view toxicity and focus their attention. Stories of flaring fill community members' narratives about Calvert City. Spectacular stories of jet-engine-like flares lighting up the night sky ultimately distract from the bioaccumulative property of dioxins handled on the complex and ongoing fugitive emissions.

Finally, this project sought to answer the question: How can understanding everyday experiences of those living in toxic places better inform our under-standing of slow violence? Stories of everyday life paint a nuanced portrait of life lived in a fenceline community. Residents mentioned being startled out of bed to the roar and glow of chemical flaring. Furthermore, there is the problem of drinking water. Despite annual water quality assessments, residents related fears of drinking the municipal water. Some cite black, oily rings appearing in the toilet and sink. As shocking as these occurrences might sound, the frequency that they occur renders them mundane. Yet, everyday knowledge about life surrounding the Calvert City Industrial Complex does not, in the eyes of the regulatory regime, warrant investigation into their claims. Without intentionally changing community narratives about industrial practices, com-munities like those surrounding the Calvert City Industrial Complex will be subjected to ongoing toxic emissions, invisible yet slowly rendering conditions for life evermore compromised.

References

Bullard, R.D. (2001), 'Environmental Justice in the 21st Century: Race Still Matters', *Phylon*, **52** (1), 72–94.

Cannuscio, C. et al. (2009), 'Visual Epidemiology: Photographs as Tools for Probing Street-level Etiologies', *Social Science & Medicine*, **69** (4), 553–564.

CDC (2018), *Facts about Benzene*, accessed 7 October 2019 at https://emergency.cdc .gov/agent/benzene/basics/facts.asp

Davies, T. (2013), 'A Visual Geography of Chernobyl: Double Exposure', *International Labor and Working-Class History*, **84**, 116–139.

Davies, T. (2018) 'Toxic Space and Time: Slow Violence, Necropolitics, and Petrochemical Pollution, Annals of the American Association of Geographers', *Annals of the American Association of Geographers*, **108** (6), 1537–1553.

de Leeuw, S. and Hawkins, H. (2017), 'Critical Geographies and Geography's Creative Re/Turn: Poetics and Practices for New Disciplinary Spaces', *Gender, Place & Culture*, **24** (3), 303–324.

Enforcement (2016), 'Case Summary: Settlement Reached for Past Response Costs at LWD, Inc. Superfund Site in Kentucky', accessed 8 July 2020 at https://www.epa .gov/enforcement/case-summary-settlement-reached-past-response-costs-lwd-inc -superfund-site-kentucky

Fritz, V. (2015), 'White supremacy fliers scattered across Marshall County', *Kentucky News Era*, accessed 5 July 2020 at https://www.kentuckynewera.com/news/article_a2e8ad24-a994-11e5-9a1a-97c009d585b6.html

Galtung, J. (1969), 'Violence, Peace, and Peace Research', *Journal of Peace Research*, **6** (3), 167–191.

Garrett, B. (2010), 'Videographic Geographies: Using Digital Video for Geographic Research', *Progress in Human Geography*, **35** (4), 521–541.

Hawkins, H. (2015), 'Creative Geographic Methods: Knowing, Representing, Intervening. On Composing Place and Page', *Cultural Geographies*, **22** (2), 247–268.

Henry, J.P. (2020), 'Spotlight: Photodrive', in Anna Feigenbaum and Aria Alamalhodaei (eds.), *The Data Storytelling Workbook*, London and New York: Routledge, pp. 235–237.

Kindon, S. (2003), 'Participatory Video in Geographic Research: A Feminist Practice of Looking?' *Area*, **35** (2), 142–153.

Lerner, Steve (2010), *Sacrifice Zones: The Front Lines of Toxic Chemical Exposure in the United States*, Cambridge, MA: The MIT Press.

Massey, Doreen (2005), *for space*, London, Thousand Oaks, New Delhi: Sage Publications Ltd.

McDonald, J. (2017), 'U.S. EPA Accelerates Progress at the BF Goodrich Superfund Site in Calvert City, Kentucky', *United States Environmental Protection Agency*, accessed 29 September 2020 at https://archive.epa.gov/epa/newsreleases/us-epa-accelerates-progress-bf-goodrich-superfund-site-calvert-city-kentucky.html

Nixon, Rob (2011), *Slow Violence and the Environmentalism of the Poor*, Cambridge, MA: Harvard University Press.

Pickett, N.R., S. Henkin, and S. O'Lear (2019), 'Science, Technology, and Society Approaches to Fieldwork in Geography', *The Professional Geographer*, **72** (2), 1–11.

Popovich, N., L. Albeck-Ripka, and K. Pierre-Louis (2020), 'The Trump Administration is Reversing 100 Environmental Rules. Here's the Full List', *The New York Times*, accessed 3 August, 2020 at https://www.nytimes.com/interactive/2020/climate/trump-environment-rollbacks.html?action=click&module=Top%20Stories&pgtype=Homepage

Pred, A. (1984), 'Place as Historically Contingent Process: Structuration and the Time-Geography of Becoming Places', *Annals of the Association of American Geographers*, **74** (2) 279–297.

Public Notice (2016), 'Draft Reissued NPDES Permit to Discharge into Waters of the State', Illinois Environmental Protection Agency Bureau of Water, Division of Water Pollution Control, Permit Section, Springfield, IL, October 26.

Release Reports (2019), United States Environmental Protection Agency, accessed 6 October 2019 at https://enviro.epa.gov/triexplorer/tri_release.chemical

Romero, A., J. Guthman, R. Galt, B. Mansfield, and S. Sawyer (2017) 'Chemical Geographies', *GeoHumanities*, **3** (1), 158–177.

Rose, Gillian (2016), *Research Methodologies: An Introduction to Researching with Visual Materials*, London: Sage.

Schwartz, Pamela M. et al. (1983), 'Lake Michigan Fish Consumption as a Source of Polychlorinated Biphenyls in Human Cord Serum, Maternal Serum, and Milk', *American Journal of Public Health*, **73** (3) 293–295.

Sturgis, S. (2012), 'Dumping Dioxin on Dixie' *Facing South*, accessed 25 September, 2020 at https://www.facingsouth.org/2012/01/dumping-dioxin-on-dixie.html

White, S. and L. Birnbaum (2009), 'An Overview of the Effects of Dioxins and Dioxin-like Compounds on Vertebrates, as Documented in Human and Ecological Epidemiology', *Journal of Environmental Science and Health*, Part C, 27 (4), 197–211.

5

Tourism development as slow violence: dispossession in Guatemala's Maya Biosphere Reserve

Jennifer A. Devine, Hannah L. Legatzke, Megan Butler and Laura Aileen Sauls

Introduction

The global tourism industry can build international solidarity, foster cultural understanding, and facilitate sustainable development (McCool and Bosak 2016). Yet, the tourism industry can also (re)create neo-colonial relations that include dispossession and violence (Devine and Ojeda 2017; Nixon 2013; Kincaid 1988). Dispossession and violence in tourism often occur swiftly and interpersonally, such as when communities lose land or usufruct rights, or in acts of sexual violence and human trafficking (Devine and Ojeda 2017). Dispossession and violence in the tourism industry, however, also unfold over longer space-time scales, transforming socio-ecological relations and the names, identities, and cultural practices of places and the people who inhabit them. Rob Nixon describes this type of "slow violence" as "a violence that is neither spectacular nor instantaneous but instead incremental, whose calamitous repercussions are postponed for years or decades or centuries" (Nixon 2013, p. 2).

This chapter explores how slow violence unfolds in Guatemala's Maya Biosphere Reserve (MBR), where the tourism industry is at the heart of a 30-year battle over access, control, and self-determination. At the center of this struggle is the El Mirador Maya archeology site near the Guatemalan-Mexican border. El Mirador's grandeur, archeological importance, and tourism potential is evident to the site's 3,000 tourists a year, including public officials, private sector developers, and conservation donors.

For 30 years, the Foundation for Anthropological Research and Environmental Studies (FARES) has preserved, excavated, and managed the El Mirador archeological site and received international accolades for its efforts (GHF 2005; FARES 2011). For much of that time, FARES's leadership has campaigned in Guatemala, and, currently, in the US Congress, to redraw the boundaries of the MBR by creating a wilderness preserve that would amplify the existing national park boundaries containing the El Mirador site. It argues for increased tourism to the site as the key generator of local economic opportunity, national economic development, and nature conservation.

At first glance, this proposal seems like an important step for wildlife conservation and cultural heritage preservation; however, its details – and particularly those captured in legislation like US Senate Bill 3131 (before the US Senate as of late 2020) – ultimately erase current land uses, livelihoods, and communities from the MBR. Many of these communities participate in community-based resource management in the Reserve. Forest concessionaries, represented by the Association of Forest Communities of Petén (ACOFOP), engage in community-based tourism and sustainably harvested timber and non-timber forest products across nine active concessions. Two forest communities have lived sustainably in this region for more than 100 years (Gómez and Méndez 2007), while Indigenous peoples claim territorial occupation since time immemorial as evidenced in El Mirador itself (Grandia 2012). Since the formation of the concession model in the mid-1990s, these communities have received international recognition for the successful conservation outcomes and social benefits generated by their community-based resource management model, including an Equator Prize and the Elinor Ostrom Prize, among others (Blackman 2015; CONAP and WCS 2018; Stoian et al. 2018).

Despite these successful conservation efforts, ecosystems and communities in the MBR continue to face threats from poaching, looting, drug trafficking, and an advancing agricultural frontier (Radachowsky et al. 2012; CONAP and WCS 2018; Devine et al. 2020). While the threats to the MBR are clear, the appropriate response to these threats are not. FARES and the community concession system represent competing visions regarding how to protect El Mirador and the entire Maya Biosphere from these threats. El Mirador is ground zero in the debate between FARES, managing the archaeological site, and proponents of the community concessions regarding which type of conservation and tourism development plan will be most effective and sustainable. The results of this long-standing legal and discursive struggle will have major implications for thousands of people currently living and working within the MBR – and the socio-ecological systems they compose.

This chapter details this struggle between FARES and the MBR's communities over tourism and development in the Reserve, including the role that geospatial technologies such as light detection and ranging (LiDAR) have played in knowledge production that facilitates slow violence in the MBR. We focus on the variegated forms of dispossession associated with tourism in this region, but also highlight the steps that communities and researchers in the MBR have taken to recognize and counteract these threats to local territorial claims and livelihoods. We conclude this chapter discussing the ways that Indigenous and peasant organizations use geospatial technologies like remote sensing and LiDAR analysis to produce knowledge about their social-environments and territorial claims.

Slow violence in tourism and geographical research methods

In evolving his analytic of slow violence, Nixon directly engages with tourism, "track[ing] the temporal and racial performances of 'wild Africa'" to demonstrate the accumulation of harm generated by the sector (2013, p. 176). Focusing on wild game reserves and hunting lodges, Nixon discusses eco-tourism's "ecologies of looking" that produce, package, and sell an imagined space – the sanitized colonial wilderness, free from human influence and yet eternally threatened by it. Such eco-tourism sells a "dominant colonial conservationist mythology" that results in a series of narratives that focus on vanishing land accompanied by vanishing wildlife and livelihoods (Nixon 2013, p. 189). The production of paradise also remakes time, taking the tourist back to a nostalgic era or pre-human nature carefully crafted for their consumption. In doing so, eco-tourism can fix people and place in a status idealized for touristic purposes – whether that means that people are fetishized or made mere labor – but divorced from the actual forces that shaped the observed landscape (Nixon 2013; Mollett 2016; Pulsipher and Holderfield 2006; Rocheleau 2015).

We use the analytic of dispossession to examine forms of violence that are not spontaneous but unfold over larger space-time scales that are relationally produced and experienced, and interconnected (Mollett 2016; Ojeda 2012; Fairhead et al. 2012). Existing literature and our own research in the Maya Biosphere suggest there are at least three categories of slow violence defining the global tourism industry. First, slow violence occurs through land dispossession in forms ranging from abrupt land grabs to incrementally lost usufruct rights. Second, slow violence operates through the commodification of culture and identity. Tourism sells the identity and experience of a place and product, which lends immense yet subtle power to tourism's representational practices that often naturalize racial difference and other forms of inequality (Devine and Ojeda 2017). Third, slow violence in tourism occurs through knowledge production about cultural heritage and tourism landscapes. Struggles over boundaries and control of tourism development at the El Mirador Maya archeology site reveal how science is marshaled to support elite land claims and limit locals' management of the Reserve's cultural and natural resources. Exposing slow violence in tourism requires critically evaluating the means of producing what Nixon (2013) describes as the "narratives about the environment" (p. 139).

This chapter builds on previous work by focusing on the role that geographic knowledge production and geospatial technologies play in (re)producing or ameliorating slow violence in tourism. We focus on geospatial technologies as a source of power to demonstrate how geographic methods enable slow violence through physical, cultural, and epistemic dispossession over time. This poses the methodological challenge of how to measure and analyze slow violence, while avoiding enacting it through knowledge production practices.

Mapping, one of the oldest geospatial technologies, has long been recognized as a performative knowledge-constructing process and a political act due to mapping's power to include and omit information, delineate boundaries, and demarcate empirical "truths" (Peluso 1995; Sletto 2009; Crampton 2009). Crampton (2009) emphasizes the performativity of mapping, describing maps not just as objects, but instead as processes that occur in "theaters for the performance and negotiation of identities" (Sletto 2009). These identities may be linked to ethnic, economic, or national visions, and maps as objects and processes may reflect imagined or ideal territorializations, as much as perceived realities. Mapping represents a negotiation of how and what geospatial information is portrayed and occurs in arenas laden with power dynamics (Bryan 2011).

Access to mapping technologies may legitimize or delegitimize claims about and to Indigenous territories (Sletto 2009; Tuhiwai Smith 2013). At the same time, we recognize that several Indigenous and peasant organizations engage with geospatial technologies, including remote sensing analysis to produce knowledge about their social-environments and territorial claims (Herlihy and Tappan 2019; Davis and Sauls 2017). These participatory and community-led mapping efforts reflect a broader push within academia to decolonize research methodologies, even given that the epistemic constraints of such ways of seeing and knowing a landscape often simplify or fix a given set of relations (Wainwright and Bryan 2009).

While mapping technologies are core to geographic research methods, qualitative methods are core to understanding people's relation to place and to contextualizing – and sometimes challenging – geospatial and cartographic imaginaries. Engaging with epistemologies that challenge the purported objectivity of positivism and technical approaches, whether to conservation or to tourism development, can bring slow violence starkly into view. As Nixon (2013) illustrates in his analysis of the safari and the game lodge, unpacking and deconstructing the epistemologies underpinning tourism development render clear the mechanisms of slow violence in tourism development.

Geographic methods, thus, hold the potential to perpetuate and/or undermine neo-colonial relations. Decolonizing geographic methods, however, is not limited to critiquing and deconstructing (neo)colonial discourses and practices in tourism development. Decolonizing methods have the potential to create collaborative research experiences that empower researchers and actors in the socio-natures of which they are part, including through participatory mapping and community-based participatory research that incorporates geospatial approaches.

To examine slow violence in tourism in the Maya Biosphere, the four North American co-authors draw on their experiences living in Reserve's forest concession communities at different times and conducting collaborative research with Reserve residents and organizations. In our projects, we have drawn on a variety of methodologies, ranging from participant observation to remote sensing. At times, we have played the role of foreign scholarly mediators interpreting local narratives and relaying our research findings in academic and non-academic output, often taking on the uneasy role of 'expert' where our community partners felt excluded. The findings and reflection in this chapter result from the generative process of comparison and mutual support over the course of ongoing research from 2009 to 2020.

Slow violence and geospatial technologies in the "heart of the Maya world"

Internationally, ACOFOP and its constituent communities are held up as a model for community-based forest management due to the positive social and environmental impacts they generate (Stoian et al. 2018). Despite these benefits, community groups legally managing forests in the Reserve face significant opposition and uncertain land-use rights. Community-based commercial timber activities are a focal point of opposition from organizations and individuals that equate forest management and timber harvesting with ecological devastation (Taylor 2010). FARES represents one such opponent of community forestry within the Reserve. Founded by archaeologist Richard Hansen, FARES's conservation efforts focus upon land that Dr. Hansen and his supporters characterize as the Mirador Basin that contradicts the existing legal boundaries of the Reserve's national parks and forest concessions (Hansen et al. 2002) (see Figure 5.1).

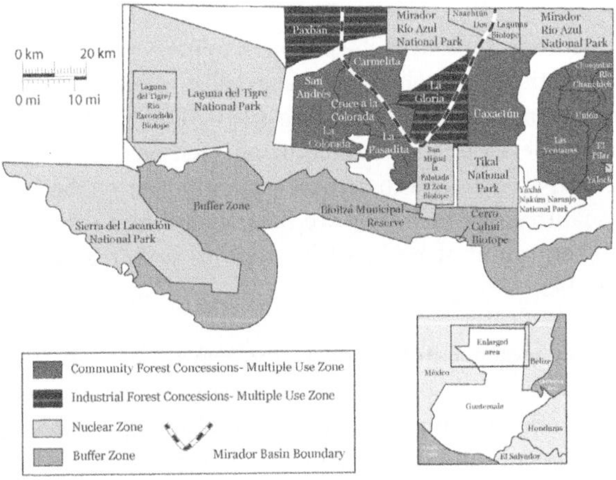

Source: Jennifer A. Devine.

Figure 5.1 The Maya Biosphere's Forest Concessions and the "Mirador Basin"

Debates about effective conservation strategies and legal protected area designations for Mirador center on whether or not the area in fact constitutes a geological basin, and both proponents of the existence or non-existence of the basin use geospatially derived knowledge to support their arguments (Rahder 2020). While FARES uses NASA Landsat satellite imagery depicting infrared signatures of vegetation types to argue that the existence of swampy, lowland forests indicates that water is pooling and the area is in fact a geologic basin, other conservation and development NGOs look to the topographic map of the region produced from data by NASA's Shuttle Radar Topography

Mission to conclude that the area does not have the characteristic slopes of a basin (Rahder 2020). Here, the designation of the area surrounding Mirador as a basin or not has implications for the political status of portions of land surrounding Mirador as either strictly protected or under community management (Rahder 2020). FARES claims that the existing boundaries of the El Mirador-Rio Azul National Park contradict the natural and cultural boundaries of the purported "Mirador Basin," the territory claimed in past and present efforts to establish a wilderness reserve where community forestry would be prohibited.

Since the early 2000s, FARES has advocated to convert parts of the Reserve currently under community management to a "wilderness area" where timber and non-timber forest product extraction would be prohibited in favor of tourism expansion. As such, the communities currently managing land within the MBR view FARES and the Mirador Basin Project as a constant threat. The community forestry system requires that tourists accessing Mirador by land contract a community tourism guide; reconfiguring national park boundaries to create a Mirador Basin wilderness area would privatize tourism services and relegate Guatemalan residents to service workers rather than protagonists of tourism development. As one resident of the Carmelita forest concession described: "I believe that there are people in the capital who are very interested in being given a concession contract for tourism management rights ... If they manage to keep tourism, I do not know what will become of all of us" (personal interview, August 7th, 2019). The region's history of land dispossession and forced migration out of protected areas/archeological sites such as Tikal and Yaxhá, as described to us by Reserve residents, amplify the concerns of community members living in the region.

Over the past two decades, FARES's efforts have periodically gained support from the US and Guatemalan governments. In 2002, the Mirador Basin Expansion Project pushed to expand the Reserve's Mirador National Park by 2,000 square kilometers to encompass the Naachatún Dos Lagunas Biotope and parts of six forest concessions. The Mirador Basin Expansion Project received substantial support from the Guatemalan president before being overturned by Guatemala's Supreme Court in 2005 after years of public protest and court cases (Cronkleton et al. 2008; Gómez and Méndez 2007; Taylor 2010).

A similar proposal, "Resolution #37 Mirador/Calakmul Cultural and Natural System," was put forth in 2010 (Wild 9 2010; Devine 2018) and again in 2016 (Bolanos 2016; Ministerio de Relaciones Exteriores 2016). Most recently, a bill entitled "Mirador-Calakmul Basin Maya Security and Conservation Partnership Act of 2019" (S.B. 3131 2019) was introduced to the US Senate by

Jim Inhofe (R-Oklahoma) and co-sponsors Tom Udall (D-New Mexico) and Jim Risch (R-Idaho). If successful, the bill could provide significant funding towards constructing infrastructure such as an eco-lodge and a train and converting the region currently under community management into a restricted access wilderness area (as detailed at www.global conservation.org; Global Conservation n.d.). The proposal, as the name suggests, also links Mirador with Calakmul in Mexico, and aspires to integrate the purported basin and wilderness area to Mexican President Andrés Manuel López Obrador's "Maya Train" megaproject, which aims to build over 900 miles of railway, creating a new tourism circuit of archaeology sites in the Yucatan peninsula (Fuentes 2018).

It is not a coincidence that the land within the Reserve that has the greatest unrealized tourism potential is also the focal point of these efforts. Tourism is a major contributor to Guatemala's GDP. The MBR's most popular archaeological site, El Tikal, draws over 250,000 visitors each year and generated nearly US\$4 million in park entrance fees alone in 2018 (Gonzales 2019). Valuable resources such as tourism potential and high-value timber often incentivize powerful actors to retain control of forest resources even in areas where community management is promoted (Anderson and MacLean 2015; Nelson and Agrawal 2008). The eco-lodge and train proposed by S.3131 would likely generate significant income for outside investors while bypassing, or even eliminating, the community-based tourism currently practiced at El Mirador. It would ultimately substantially restrict the access of local communities to the land they have historically managed.

This ongoing struggle by communities to live in and manage tourism and other resources in the MBR illustrates the challenges and opportunities associated with the role of science and knowledge production in environmental management. Rather than being an impartial tool, scientific inquiry can be utilized to highlight or question the legitimacy of current forest management practices (Anderson and MacLean 2015). Scientific inquiry can also be utilized to justify competing approaches to forest management. In such "adversarial science" (Clapp and Mortenson 2011), scientific inquiry is utilized as a weapon and tool in debates utilized to discredit opponents and promote a specific narrative/plan of action. In the MBR, mapping has been a powerful weapon. The forests of the Maya Biosphere Reserve, sharing international boundaries with Belize and Mexico, acutely exemplify the political struggle for territory through mapping and remote sensing technologies, especially in light of political rhetoric that depicts forest dwelling communities as invaders.

Claiming the MBR for pristine wilderness and an idealized Maya past, in spite of the living Maya and their descendants' ongoing presence on and relations with the land, enacts an ongoing form of slow violence through tourism imaginaries. In FARES's mapping practices, land currently under community management is not acknowledged. FARES's description of the region portrays it as a virgin forest devoid of human influence, a "system [that] spreads across 1.6 million acres of pristine tropical forest" (FARES Foundation 2018). This narrative presents the Mirador Basin as an ever-diminishing oasis of forested area surrounded by threats. The organization's website reads, "The forests in this region are highly threatened, but can provide new economic benefits for communities and the Republic of Guatemala through the establishment of world class archaeological and natural preserves" (FARES Foundation 2018). This narrative ignores local communities' history in the forest as well as their role in forest protection. This vision of the Reserve is then utilized to posit conservation solutions that leave local communities out of the equation.

In contrast, other recent studies utilizing remote sensing technologies have broken down the Reserve to identify the areas currently under forest management. These studies paint a very different picture and acknowledge the effectiveness of community forest management at preventing deforestation (Hodgdon et al. 2015; Blackman 2015; CONAP and WCS 2018). These efforts to re-delineate the boundaries within the MBR are executed in the name of archaeological and environmental conservation by different actors seeking to control and utilize the land in the service of different ideas about appropriate use. Boundary making in archeology and conservation projects is a means of creating and controlling tourism concessions, just as the concessions' boundary making seeks to legitimize their role as conservation and development actors.

Emerging technologies enable ever more 'scientific' ways to engage in delineation, boundary making, and legitimization. The control and application of LiDAR, especially, has powerful implications for who is involved in the reproduction of spatial knowledge, and how narratives of space are portrayed. LiDAR technologies emit pulses of electromagnetic radiation and measure the time it takes emitted light to return to the sensor. Aircraft equipped with LiDAR can thereby create high resolution and multi-dimensional models of forest structure, topography, and landscape features, such as archaeological sites under a forest canopy, when flown over an area (Fernandez-Diaz et al. 2014; Chase et al. 2012). In 2016, a Guatemalan non-profit, the Pacunam Foundation for Maya Cultural and Natural Heritage, began a three-year initiative to map the Guatemalan lowlands using LiDAR. The project is the largest LiDAR survey to date, covering 2144 km2 of the Maya Biosphere Reserve

(Canuto et al. 2018; Clynes 2018). This aerial survey covers greater area than possible through on-the-ground archaeological research in the tropical forested area and detects topography and features of archaeological sites through the scanning of the terrain from six view angles and higher sensor-range resolution than used previously (Canuto et al. 2018).

The risk with LiDAR lies not within the technology itself, but the disproportionate access to these technologically sophisticated means of producing spatial information. The use of LiDAR in the MBR becomes a means of slow violence when marshalled in practices of scientific knowledge production that inject new life into neo-colonial narratives of discovery. LiDAR allows scientists to gather information about archaeological structures beneath a forest that makes such areas less accessible to on-the-ground investigation, yet residents of the MBR have known of these sites' existence for decades, taking archeologists to the sites who then claim to have 'discovered them.' Media coverage of the MBR LiDAR campaign, as well as the narratives FARES and PACUNAM construct and circulate, present the MBR's sites as lost and unknown, and the application of LiDAR as bringing what was once invisible to light (i.e. Cascone 2019; Clynes 2018). However, in making discoverable these archaeological sites, and presenting the application of LiDAR as apolitical and objective, this coverage simultaneously invisibilizes the living human communities of the MBR and their contribution to archeological knowledge production and conservation efforts, rendering them in Nixon's words "unimagined communities" (2013, p. 150).

Tourism marketing materials by FARES, PACUNAM, and the Guatemalan Institute of Tourism promote imagery of El Mirador and the MBR as a timeless landscape of lost civilizations devoid of contemporary residents, while coverage of LiDAR-derived archaeological findings reproduces and reinforces this imagery, mentioning the modern-day presence of human communities in the region only to discuss "modern-day looters" and "trespassers" who "burn and clear land for agriculture and human settlement" (Clynes 2018). The erasure of communities through narratives of scientific discovery builds on and articulates with depictions of the MBR as pristine wilderness in need of protection.

We identify this slow violence in tourism development and cultural heritage conservation as a form of epistemic dispossession. While local communities increasingly use remote sensing technologies in their own forestry management and mapping, the disproportionate access to these technologies and propagation of LiDAR-derived discovery narratives contributes to the erasure of local territorial and cultural knowledge claims. As one member of a community forestry concession described in reference to these studies, "People have even arrived who have done studies, but we do not know what they study. Far from feeling content we feel worried about what they will do and say. Like the application of the lasers. We already knew that information" (personal interview, April 1st, 2018).

The technological fetishism of promoting and distributing LiDAR-based knowledge ignores the spatial-temporal narratives of MBR communities, who in many cases already understand the archeological landscape they inhabit through their careful, holistic management of these forest spaces. In constructing a narrative of discovery through omission, media coverage of LiDAR mapping contributes to the erasure of local struggles for territory. As one member of ACOFOP describes, "Have you seen about the LiDAR? And the National Geographic program? The majority of the [archaeological sites] that appear in these programs are sites within the community concessions. They never mention that the sites are there because of the communities ... they look to invisibilize the work done by the communities and promote their model" (personal interview, June 20th, 2019; see also Devine 2016).

The mounting use of geospatial technologies by ACOFOP suggests a recognition of the power of these technologies to make claims about territory seem more rational and scientifically sound (Davis and Sauls 2017; Rahder 2020). Currently, the community monitoring network of ACOFOP uses drones and geographic information systems for fire-prevention and strategic planning in the concessions (ACOFOP 2020). ACOFOP uses these geospatial technologies to document successful community concession management, and has displayed drone imagery of the Carmelita concession on a TV Indigena (online) broadcast to critique the omission of communities from S.3131 bill proposing to expand and privatize tourism around El Mirador (TV Indigena 2020).

Access to spatial technologies enables the propagation of narratives and counternarratives of forested spaces. Disproportionate access to these technologies, however, creates power imbalances. Their use also perpetuates Western, positivist epistemologies and their disciplinary technologies of defining what information and experience counts as legitimate knowledge about territorial claims. Community foresters are acutely aware of these inequalities, as an ACOFOP representative explained: "In our country, what a foreigner says matters more than what Guatemalans themselves say" (personal interview, June 20th, 2019). Ownership of this technology, proficiency in English, and 'gringo' identity combine to empower certain narratives told about the Reserve and its inhabitants.

In the face of S. 3131, community foresters and their allies, once again, have invested massive amounts of money, time, and political capital to defend their land rights and livelihoods. Marshaling another technology, social media, supporters of the concessions and opponents of S. 3131 created a Change. org petition, 'Stop bill S.3131 – Mirador-Calakmul Basin Maya Security and Conservation Partnership Act,' that had been signed by more than 200,000 people as of September 2020. As supporters of the community forestry system in the MBR, we are concerned about how this bill will embolden FARES's vision and territorial claims, and enact direct dispossession and perpetuate slow violence through tourism development in the region.

Conclusion

This chapter has focused primarily on the last 30 years of territorial and discursive battle over the Maya Biosphere Reserve. Yet, this latest articulation of slow violence in tourism development builds upon a much longer 500-year history of land dispossession of Indigenous peoples in the Americas (Grandia 2012). We focus on this recent history to illuminate how the use of contemporary geospatial technologies in archeological research articulates with living histories of colonialism to create practices and processes of slow violence in tourism development.

Slow violence is structural and spatial and unfolds across time and space in ways that can make it less visible than acts of interpersonal violence; slow violence is epistemic and, thus, enables and justifies other forms of violence (Nixon 2013). Epistemic violence through geospatial knowledge production about archeological sites both enacts dispossession through the erasure of non-Western ways of knowing and enables other forms of violence such as land dispossession and cultural appropriation. Rob Nixon (2013) explains slow violence in terms of how tourism destinations, like the African Safari, create anachronistic spaces that reproduce colonial encounters and power relations. We build on these insights and our collective experiences in Guatemala's Maya Biosphere Reserve to argue that geospatial technologies facilitate and enable multiple forms of dispossession, and the reproduction of neo-colonial power relations, narratives, and identities in tourism. The Maya Biosphere Reserve suggests that these profound transformations occur before the tourists even arrive and are in fact essential for large-scale tourism to exist.

Geospatial technologies and methods are marshaled in contemporary legal and discursive debates about the future of tourism development and conservation of the Maya Biosphere. For more than 30 years, the North American-led research organization FARES and its supporters have used remote sensing and other geospatial technologies, such as remote sensing and LiDAR, to define and demarcate a geography it has named the Mirador Basin. FARES argues that the physical and cultural integrity of the basin is incongruent with the legal boundaries existing in the Maya Biosphere that demarcate a much smaller area as the El Mirador-Rio Azul National Park. However, the science is anything but solid, and the existence of the Maya Basin is contested (Devine 2018; Rahder 2020). Nonetheless, FARES and the organization's supporters have used their contested scientific claims to campaign the Guatemalan and US governments to reconfigure the boundaries and territoriality of the Maya Biosphere to transform the geographical imaginary of the Mirador Basin into a legal territory.

In this context, spatial information becomes woven into political struggles for territory. S. 3131, introduced to the US Senate in December 2019, states that the sustainability of the conservation plan depends on the privatization and development of eco- and archeological tourism services. Supporters of the 'Mirador Basin' and S.3131 leverage purported 'discoveries' made possible through LiDAR and present them as scientific fact to inform legislation with potentially dramatic consequences for the management of the MBR.

This geographic imaginary threatens the integrity of grassroots community forestry because the proposed expanded legal boundaries of the Mirador basin and wilderness area will reduce the territory of four of nine operating forest concessions. Furthermore, many Reserve residents and supporters feel that the initiative threatens the integrity of the community-based resource management as a whole. For the last 20 years, the umbrella organization of ACOFOP has expended scarce resources and political capital to defend the successful conservation model they fought so hard to build (Devine 2018). The community forestry system enables community foresters to be protagonists in future development of the reserve, rather than simply providing labor to tourism businesses owned and operated by Guatemalan or North American elites.

Preventing slow violence in tourism, particularly in formerly colonized countries, requires attention to and respect for the experiences and knowledges of those living in and around protected areas. These peoples and communities argue that they are and should be considered protagonists in the story of conservation and economic development in the landscapes in which they live. In the MBR, they have proven themselves – using scientific metrics as well as their own ways of knowing – as both successful guardians of their forest and as generators of community well-being. In the current era of green neoliberalism, however, we must look outside of the Reserve to understand drivers of slow violence in Mirador tourism development, and we must ask ourselves as researchers how novel geospatial technologies inject new life into enduring, living histories of colonialism and dispossession in its diversity of pernicious forms.

References

Anderson, W. F. and D. A. MacLean (2015), 'Public forest policy development in New Brunswick, Canada: multiple streams approach, advocacy coalition framework, and the role of science', *Ecology and Society*, **20** (4), 20.

Association of Petén's Forest Communities (ACOFOP) (2020), *Monitoring Network*, accessed 25 September 2020 at https://acofop.org/en/mujeres-y-jovenes/#monitoring_network

Blackman, A. (2015), 'Strict versus mixed-use protected areas: Guatemala's Maya Biosphere Reserve', *Ecological Economics*, **112**, 14–24.

Bolanos, R. (2016), 'Proponen instalar un tren hacia El Mirador', *Prensa Libre*, accessed 14 October 2020 at https://www.prensalibre.com/economia/proponen-instalar-un-tren-hacia-el-mirador/

Bryan, J. (2011), 'Walking the line: participatory mapping, indigenous rights, and neo-liberalism', *Geoforum*, **42** (1), 40–50.

Canuto, M. A., F. Estrada-Belli, T. G. Garrison, S. D. Houston, M. J. Acuña, M. Kováč and D. Chatelain (2018), 'Ancient lowland Maya complexity as revealed by airborne laser scanning of northern Guatemala', *Science*, **361** (6409).

Cascone, S. (2019), 'How lasers are utterly transforming our understanding of the ancient Maya, bringing their whole civilization back to light', *Art World*, 8 August, accessed 25 September 2020 at https://news.artnet.com/art-world/technology-transforming-mayan-archaeology-1558456

Chase, A. F., D. Z. Chase, C. T. Fisher, S. J. Leisz and J. F. Weishampel (2012), 'Geospatial revolution and remote sensing LiDAR in Mesoamerican archaeology', *Proceedings of the National Academy of Sciences*, **109** (32), 12916–12921.

Clapp, R. and C. Mortenson (2011) 'Adversarial science: conflict resolution and scientific review in British Columbia's central coast', *Society and Natural Resources*, **24** (9), 902–916.

Clynes, T. (2018), 'Exclusive: laser scans reveal Maya "megalopolis" below Guatemalan jungle', *National Geographic*, 8 February, accessed 24 September 2020 at https://www.nationalgeographic.com/news/2018/02/maya-laser-lidar-guatemala-pacunam/

Consejo Nacional de Áreas Protegidas (CONAP) and Wildlife Conservation Society (WCS) (2018), Monitoreo de la gobernabilidad de la Reserva Biosfera Maya, accessed 14 October 2020 at Monitoreo de la gobernabilidad de la Reserva Biosfera Maya.

Crampton, J. W. (2009), 'Cartography: performative, participatory, political', *Progress in Human Geography*, **33** (6), 840–848.

Cronkleton, P., P. L. Taylor, D. Barry, S. Stone-Jovicich and M. Schmink (2008), 'Environmental governance and the emergence of forest-based social movements', CIFOR Occasional Paper 49.

Davis, A. and L. Sauls (2017), 'Evaluating forest fire control and prevention effectiveness in the Maya Biosphere Reserve', ACOFOP and PRISMA, accessed 21 October

2020 at www.prisma.org.sv/wp-content/uploads/2020/01/evaluating_forest_fire
_MBR.pdf

Devine, J. A. (2016), 'Contesting global heritage in the *Chicle* worker's museum', *Latin America Research Review*, **51** (3), 101–122.

Devine, J. A. (2018), 'Community forest concessionaires: resisting green grabs and producing political subjects in Guatemala', *The Journal of Peasant Studies*, **45** (3), 565–584.

Devine, J. and D. Ojeda (2017), 'Violence and dispossession in tourism development: a critical geographical approach', *Journal of Sustainable Tourism*, **25** (5), 605–617.

Devine, J. A., N. Currit, Y. Reygadas, L. I. Liller and G. Allen (2020), 'Drug trafficking, cattle ranching and land use and land cover change in Guatemala's Maya biosphere reserve', *Land Use Policy*, **95**, 104578.

Fairhead, J., M. Leach and I. Scoones (2012), 'Green grabbing: a new appropriation of nature?', *The Journal of Peasant Studies*, **39** (2), 237–261.

FARES (2011), *Foundation for Anthropological Research and Environmental Studies*, accessed 22 September 2020 at https://www.fares-foundation.org/mirador/archaeology.php

FARES Foundation (2018), '"Where and What?" Mirador Basin Project', accessed 21 October 2020 at www.miradorbasin.com/where-and-what/

Fernandez-Diaz, J., W. Carter, R. Shrestha and C. Glennie (2014), 'Now you see it … Now you don't: understanding airborne mapping LiDAR collection and data product generation for archaeological research in Mesoamerica', *Remote Sensing*, **6** (10), 9951–10001.

Fuentes, Y. (2018), 'Tren Maya: así es el ambicioso proyecto que propone AMLO y tiene un costo de miles de millones de dólares para México', *BBC News Mundo*, 15 November, accessed 14 October 2020 at https://www.bbc.com/mundo/noticias -america-latina-45254080

GHF Public Relations (2005), 'GHF Leader in Conservation Dr. Richard Hansen Awarded National Order of the Cultural Patrimony of Guatemala', *Global Heritage Fund*, accessed 14 October 2020 at https://globalheritagefund.org/news/ghf-leader -in-conservation-dr-richard-hansen-awarded-national-order-of-the-cultural -patrimony-of-guatemala/

Global Conservation (n.d.), 'Mirador National Park, Guatemala', accessed 5 October 2020 at https://globalconservation.org/projects/mirador-national-park-peten -guatemala-1/

Gómez, I. and V .E. Méndez (2007), *Association of Forest Communities of Petén, Guatemala Context, Accomplishments and Challenges*, Bogor Barat 16680, Indonesia: CIFOR, accessed 14 October 2020 at www.cifor.org/knowledge/publication/2464/

Gonzales, P. (2019), 'Gobierno recaudó Q29.6 millones por cobro bancarizado en el Parque Nacional Tikal', *Diario de Centro América*, accessed 28 September 2020 at https://dca.gob.gt/noticias-guatemala-diario-centro-america/gobierno-recaudo-q29 -6-millones-por-cobro-bancarizado-en-el-parque-nacional-tikal/

Grandia, Liza (2012), *Enclosed: Conservation, Cattle, and Commerce Among the Q'eqchi' Maya lowlanders*, Seattle: University of Washington Press.

Hansen, R. D., S. Bozarth, J. Jacob, D. Wahl and T. Schreiner (2002), 'Climatic and environmental variability in the rise of Maya civilization: a preliminary perspective from northern Petén', *Ancient Mesoamerica*, **13** (2), 273–295.

Herlihy, P. H. and T. A. Tappan (2019), 'Recognizing indigenous Miskitu territory in Honduras', *Geographical Review*, **109** (1), 67–86.

Hodgdon, B., D. Hughell, V. H. Ramos and R. McNab (2015), 'Deforestation trends in the Maya biosphere reserve, Guatemala', New York: Rainforest Alliance, accessed

15 October 2020 at https:www.rainforest-alliance.org/impact-studies/deforestation
-trends-maya-biosphere-reserve

Kincaid, Jamaica (1988), *A Small Place*, New York: Macmillan.

McCool, Stephen F. and Keith Bosak (eds.) (2016), *Reframing Sustainable Tourism*, Dordrecht, Netherlands: Springer Netherlands.

Ministerio de Relaciones Exteriores (2016), 'Acuerdo Gubernativo Número 211-2016, accessed 14 October 2020 at https://sgp.gob.gt/wp-content/uploads/2016/11/AG -211-2016-1.pdf

Mollett, S. (2016), 'The power to plunder: rethinking land grabbing in Latin America', *Antipode*, **48** (2), 412–432.

Nelson, F. and A. Agrawal (2008), 'Patronage or participation? Community-based natural resource management reform in sub-Saharan Africa', *Development and Change*, **39** (4), 557–585.

Nixon, Rob (2013), *Slow Violence and the Environmentalism of the Poor*, Cambridge, MA: Harvard University Press.

Ojeda, D. (2012), 'Green pretexts: ecotourism, neoliberal conservation and land grabbing in Tayrona National Natural Park, Colombia', *Journal of Peasant Studies*, **39** (2), 357–375.

Peluso, N. L. (1995), 'Whose woods are these? Counter-mapping forest territories in Kalimantan, Indonesia', *Antipode*, **27** (4), 383–406.

Pulsipher, Lydia M. and Lindsey C. Holderfield (2006), 'Cruise tourism in the Eastern Caribbean: an anachronism in the post-colonial era?', in Ross Kingston Dowling (ed.) (2006), *Cruise Ship Tourism*, Oxford, UK and Cambridge, MA: CABI.

Radachowsky, J., V. H. Ramos, R. McNab, E. H. Baur, and N. Kazakov (2012), 'Forest concessions in the Maya biosphere reserve, Guatemala: a decade later', *Forest Ecology and Management*, **268**, 18–28.

Rahder, Micha (2020), *An Ecology of Knowledges: Fear, Love, and Technoscience in Guatemalan Forest Conservation*, Durham, NC: Duke University Press.

Rocheleau, D. E. (2015) 'Networked, rooted and territorial: green grabbing and resistance in Chiapas', *The Journal of Peasant Studies*, **42** (3–4), 695–723.

S.B. 3131 (2019), Mirador-Calakmul Basin Maya Security and Conservation Partnership Act of 2019, 116th Congress (2019), accessed 14 October 2020 at http:www.congress .gov/bill/sentate-bill/3131/text

Sletto, B. I. (2009), '"We drew what we imagined" participatory mapping, performance, and the arts of landscape making', *Current Anthropology*, **50** (4), 443–476.

Stoian, D., A. Rodas, M. Butler, I. Monterroso and B. Hodgdon (2018), 'Forest concessions in Petén, Guatemala: a systematic analysis of the socioeconomic performance of community enterprises in the Maya Biosphere Reserve', CIFOR, accessed 14 October 2020 at http://www.cifor.org.knowledge/publication/7163

Taylor, P. L. (2010), 'Conservation, community, and culture? New organizational challenges of community forest concessions in the Maya Biosphere Reserve of Guatemala', *Journal of Rural Studies*, **26** (2), 173–184.

Tuhiwai Smith, Linda (2013), *Decolonizing Methodologies: Research and Indigenous Peoples*, London: Zed Books Ltd.

TV Indigena (2020), 'Salvemos el mirador', *TV Indigena*, accessed 24 July 2020 at https://www.facebook.com/1859759450905583/videos/2608972479417815

Wainwright, J. and J. Bryan (2009), 'Cartography, territory, property: postcolonial reflections on indigenous counter-mapping in Nicaragua and Belize', *Cultural Geographies*, **16** (2), 153–178.

Wild 9: Ninth World Wilderness Congress (2010), 'Resolution # 37: conservation of the Mirador/Calakmul cultural and natural system', accessed 14 October 2020 at https:// www.wild.org/wp-content/uploads/2009/11/Res37_Eng.pdf

6 From violent conflict to slow violence: climate change and post-conflict recovery in Karamoja, Uganda

Daniel Abrahams

Introduction[1]

Academic inquiry regarding climate change and security has largely centered on understanding whether, how, and to what degree, climate change does or does not alter conflict outcomes (Abrahams & Carr, 2017; Ide, 2017). Many critical scholars, however, have argued that the focus on climate change as a factor of conflict risks overlooking the human security implications of climate change or environmental change more generally (see, for example, Barnett, 2001, 2019; Gemenne et al., 2014; Oels, 2015; Verhoeven, 2011). This chapter employs Rob Nixon's slow violence framework, which emphasizes the incremental and accretive repercussions of environmental change while reconceptualizing environmental change as a phenomena that does not simply drive violence, but imparts it (Nixon, 2011). I employ this framework to examine how Karamoja, Uganda – a region recovering from years of intense violent conflict – experiences the impacts of climate change as a form of slow violence.

There has been only limited research that utilizes the slow violence framework in analyzing the drivers and impacts of climate change (O'Lear, 2016; Willett, 2015). However, scholars from across multiple disciplines have employed Johan Galtung's (1969) structural violence framework to articulate the myriad ways in which the poor – largely in the global south – experience the structural inequalities of the global economic system and concomitant impacts of environmental degradation (see, for example, Kaufman, 2014; O'Lear, 2016; Wapner, 2014; Watts, 1983). In describing structural violence, Galtung demonstrates that there are multiple dimensions of violence and that violence might not be imparted directly by one person onto another (Galtung, 1969). Nixon's slow violence thesis directly builds on Galtung's work on structural violence; however, where it differs, and as it is employed in this chapter, is in its explicit emphasis on the insidious and invisible impacts of environmental change.

In examining how climate change manifests as a form of slow violence, I draw upon not only Nixon's framework, but also the political ecology, science and technology studies (STS), and environmental security literatures that critically examine the relationship between environmental change and security outcomes (see, for example, Barnett, 2001; Dalby, 2009; Le Billon, 2001; Le Billon & Duffy, 2018; O'Lear & Dalby, 2015; Peluso & Watts, 2001). In so doing, my approach to understanding violence expands beyond binary states of conflict and peace, and rejects the notion that those who impart violence and those who suffer are spatially or temporally connected, or that conflict need be a physical outcome (see, for example, Barnett, 2001; Barnett & Adger, 2007; Branch, 2018; Detges, 2014; Galtung, 1969; Hecht, 2018; Ide et al., 2014; Verhoeven, 2011; Wapner & Elver, 2016).

The epistemologies of slow violence align closely with the political ecology and STS literatures that examines the overlapping challenges of climate change and conflict (climate-conflict) (e.g., Le Billon, 2001; Peluso, 2008; Peluso & Watts, 2001; Turner, 2004; Watts, 2004). Like the structural violence and slow violence frameworks, these bodies of literature demonstrate that violence is embedded in power and social structures as well as the discourses that reinforce those structures and power imbalances (e.g., Okpara et al., 2016; Selby & Hoffmann, 2014). Likewise, both STS and political ecology emphasize the importance of scale in understanding environmental change and violent conflict (e.g., O'Lear, 2005; O'Loughlin et al., 2012), connecting global political economic processes with localized outcomes (Hecht, 2018; Verhoeven, 2011), and deconstructing the discourses of conflict and environmental change (e.g., Branch, 2018; Floyd, 2008; Verhoeven, 2011). All of these schools of thought emphasize that the role environmental change plays in conflict is neither apolitical nor acontextual. Indeed, despite ongoing debate regarding the relationship between climate change and conflict, a central conclusion in the climate-conflict literature is that climate change's relationship to conflict is neither direct, nor is it as impactful as socioeconomic and other structural vulnerabilities (Mach et al., 2019). As I demonstrate in this chapter, the slow violence of climate change, too, is always part of a broader set of drivers of vulnerability.

Karamoja, Uganda: intersecting challenges of climatic variability and conflict recovery

Karamoja is the northeastern-most region of Uganda. It is a vast area, roughly the size of Rwanda, with a population of roughly one million people (Mercy Corps, 2016; Nakalembe et al., 2017). Karamoja's patterns of rainfall and aridity are central factors in its political ecology. The rainfall is unimodal (i.e., there is a one rainy season per year) with most areas in Karamoja receiving between 500–1,000 mm per year (Chaplin et al., 2017; Mubiru, 2010; Nakalembe et al., 2017). As such, the region has historically relied on livelihood systems that utilized mobility as a hedge against this variability – namely agropastoralism and pastoralism (Bushby & Stites, 2016; Mamdani, 1982; Nakalembe et al., 2017).

Karamoja is recovering from years of violent conflict. These conflicts were ostensibly driven by cattle raids between rivalrous and ethnically defined pastoralist groups (Eaton, 2007, 2008; Howe et al., 2015; O'Keefe, 2010). There was, however, no single cause of this violence. Indeed, reflecting the misconception that the violence experienced from the early-1980s through the mid-2000s was simply the result of cattle raids, raids had long existed prior to this period but were closely governed by elders, conducted primarily with bows, arrows, and spears, and fatalities were rare. These raids served multiple purposes including cattle restocking, territorial exchange, and wealth accumulation (Filipová & Johanisova, 2017; O'Keefe, 2010). However, following an influx of small arms – many of which were acquired following the fall of Idi Amin's government in 1979 – increasing degradation of grazing land, and the commercialization of cattle raids, the scope of raids intensified, and the ability of elders to govern raids diminished. As the conflict grew both in spatial scale and in scope, retaliations and counter-retaliations became common (Eaton, 2007, 2008; Howe et al., 2015; Mkutu, 2010; Stark, 2011). According to Gray et al. (2003), more than 70 percent of deaths for males, aged 30–39 in the sub-tribes she interviewed (Bokora and Maheniko) were the result of cattle raiding.

In 2006, the central government imposed a disarmament campaign under the auspices of the Karamoja Integrated Disarmament and Development Programme (KIDDP). This disarmament process was incredibly violent and associated with widespread human rights violations. However, it was seemingly successful in its intended outcomes (Human Rights Watch, 2007; Stites & Akabwai, 2010; UNHCHR, 2007). Following the disarmament, and with continued presence of the Uganda People's Defence Force (UPDF) and police forces intended to limit cattle raids, Karamoja stabilized. Due to these shifts, contemporary Karamoja can also be understood as being in a 'post-conflict' period. Although this distinction is difficult to define in concrete spatial or temporal terms, Karamoja's challenges in recovering economically following the prolonged conflict and the challenges in avoiding conflict recurrence are typical of the fragility of a place recovering from conflict (Collier et al., 2008).

Karamoja has not experienced a seamless transition to peace and prosperity (Abrahams, 2020). Rather, this post-conflict period, or what many survey respondents and interview participants describe as a 'period of development,' has been marked by new vulnerabilities driven by the intersecting challenges of region-wide changes in land use and land tenure policy, intensifying climatic variability, and the intersection of those issues. These compounding factors have led to increasing rates of localized, small-scale conflict and new vulnerabilities to the more extreme variability and warmer temperatures. Though there has been a steady decrease in the quantity and severity of violent conflict as compared to 'those days' – a common term describing the years with high rates of violent conflict – in combination with region-wide shifts in the wider political economy, the effects of climate change have driven new insidious forms of vulnerability. In other words, Karamoja has shifted from experiencing high rates of severe violent conflict, to the insidious and accretive impacts of slow violence.

Seeing the slow violence of climate change

Violence is generally conceived as a time-bound event or action that is explosive, dramatic, and, therefore, apparent (Galtung, 1969). Conversely, the impacts of environmental degradation and pollution are latent, slow moving, and disconnected from its original cause – both in space and time (Nixon, 2011; O'Lear, 2016). Drawing upon Peluso and Watts's (2001) work on violent geographies, Nixon (p. 8) situates the ontology of environmental degradation as slow violence as follows:

> Violence, above all, environmental violence, needs to be seen – and deeply considered – as a contest not only over space or bodies, or labor, or resources, but also over time.

Environmental change, and in particular global climate change, is reflective of Nixon's description of slow violence as something that transcends physical violence to encompass temporality as well. Greenhouse gases (GHGs) enter the atmosphere gradually, and with no visible pollution, the impacts of climate change are defined largely in probabilistic terms, and the impacts are not confined to where – or when – most GHGs are emitted into the atmosphere.

Moreover, those most acutely affected by climate change are those least responsible for GHG emissions (Ravallion, 2000). The practical implications of this invisibility are severe. Attention, and policy, are drawn to the dramatic and newsworthy as opposed to the slow moving and quotidian impacts of pollution and environmental degradation – what Nixon calls an imaginative dilemma (Nixon, 2011, pp. 3, 11). This invisibility also raises a difficult epistemological question: If the defining feature of slow violence is its lack of visibility, how can one make visible the slow violence of climate change?

This chapter draws upon research and data collection initially guided by the goal of understanding the ways in which climate change alters conflict outcomes in Karamoja, Uganda, and how development organizations seek to address that potential link (Abrahams, 2018). Research design was informed by political ecological framings of climate-conflict, which argues that climate change's impact on social, economic, and security outcomes cannot be extricated from the context of place (Abrahams & Carr, 2017; Le Billon & Duffy, 2018). As such, though this study centered on climate change and conflict, data collection also addressed topics related to historical conflict, development, and economic change in the region, shifts in land use and land policy, the impacts of extractive industry, and myriad other topics that constitute Karamoja's political ecology. Therefore, this research design and the data collection it involved are well-suited to an examination of the contextual characteristics of slow violence.

I utilized a mixed-methods approach to data collection, which focused on triangulation to increase validity (i.e., utilizing multiple methods to gain a broader perspective of and better understand the case) (Maxwell, 1992). The primary data source informing this study is a survey of local governance officials (n=103). The sample was most heavily represented by local government officials in Karamoja (n=78). The plurality of the sample (n=45) are Local Council 1s (or LC1), representatives in the smallest administrative unit in the country (for more detail on survey composition see Abrahams, 2018). This survey was designed to elicit both quantitative and qualitative responses. For example, using benchmarked timeframes, I asked participants to rank the severity of conflict and climate change on a Likert scale, in order to understand the shifts in the region over time. These questions allowed for follow-up and clarifying questions which provided key contextual understandings of the quantitative data. In response, research participants provided nuanced answers on how the changes being experienced in Karamoja are felt differently across the population.

This research also draws upon six months of participant observation in Karamoja, select interviews with key informants (n=5), a systematic review of the gray literature, and a presentation of preliminary findings to research participants as a means to validate the data (for a more detailed description of these particular methods, see Abrahams, 2018). The participant observation focused on development organizations' efforts related to climate-conflict in Karamoja, Uganda. Through my participation in the day-to-day activities of a particular development organization, Mercy Corps, I was able to take part in activities with communities such as resilience assessments, meetings with local government officials, and program evaluations as well as Mercy Corps strategy sessions in which they developed programmatic plans. Critically, this involvement also enabled me to have discussions with research participants regarding climate change, conflict, and economic changes in Karamoja. Though I do not classify these conversations as interviews due to their informal character, by situating questions based on the immediate contexts, these conversations proved to be a valuable data source for understanding socioecological changes in Karamoja. All observational data were recorded utilizing detailed fieldnotes (Emerson et al., 2011).

Though informed by a mixed-methods approach, I draw heavily upon participants' qualitative descriptions of climate change and conflict in this analysis. This approach enabled not only an interrogation of climate change's role as a contributing factor on acute conflict outcomes, but also a better situation of climate change into Karamoja's wider political ecology. Moreover, by separately examining the impacts of climate change and types of conflict in the region, as well as individuals' descriptions of both climate change and conflict, I was able to demonstrate why participants' descriptions of climate change closely mirrored their descriptions of conflict. In concert with other methods, this emphasis on the rhetoric employed by research participants was critical in contextualizing the difficult-to-identify violence imparted by the effects of climate change.

From widespread violence to slow violence: a post-conflict paradox

Climate change acts as a form of slow violence in Karamoja multiple, overlapping ways. First, in combination with wider political ecological changes, climate change has been a contributing factor to localized forms of conflict and risks the current stability in the region. Second, these shifts have created new forms of vulnerability that alter individuals' livelihoods, agency, and identity. I examine these overlapping impacts in the following subsections.

Climate change, post-conflict fragility, and localized conflict outcomes

Research participants consistently described how the impacts of climate change drove conflict outcomes and negatively affected the region's current stability. They most often described this as a dry season challenge. For example, in response to a question about the most pressing security concerns in Karamoja, a security official from Tapac subcounty noted that conflicts tend to increase in the dry season. This was a commonly referenced pattern. As he explained:

> During October to December conflicts tend to increase, this is during dry season when many people look for several ways to survive.

That pattern is reflected in the quantitative survey data. Though the populations' perception of overall severity of conflict is clearly decreasing in Karamoja (see Figure 6.1), the perceived difference between the severity of conflicts in the dry season and rainy season is increasing. The data demonstrate that at $p<.01$ there is no statistically significant difference between conflict in the dry season and conflict in the rainy season during the 1997–2007 time period. But, at $p<.01$ there is a statistically significant difference between the seasons in all other time periods. There is also a statistically significant difference at $p<.01$ of the overall changes between conflict in the dry season and conflict in the rainy season over time. That is, while there are no clear environmental linkages to the more severe conflict of 'those days,' there does appear to be an environmental linkage to the more localized and less severe conflict in present-day Karamoja (see also Eaton, 2007, 2008).

Aligning with the quantitative data, the qualitative descriptions of climate-conflict consistently followed a particular pattern: participants described longer, hotter dry seasons, unpredictable rainfall patterns, and, in some instances, extreme precipitation. Those changes contributed to seasonal migration, a movement of livestock, and, in turn, negative outcomes to their livelihoods. Some of these negative outcomes contributed to increasing rates of small-scale conflict between neighbors or neighboring communities, theft, and domestic violence.

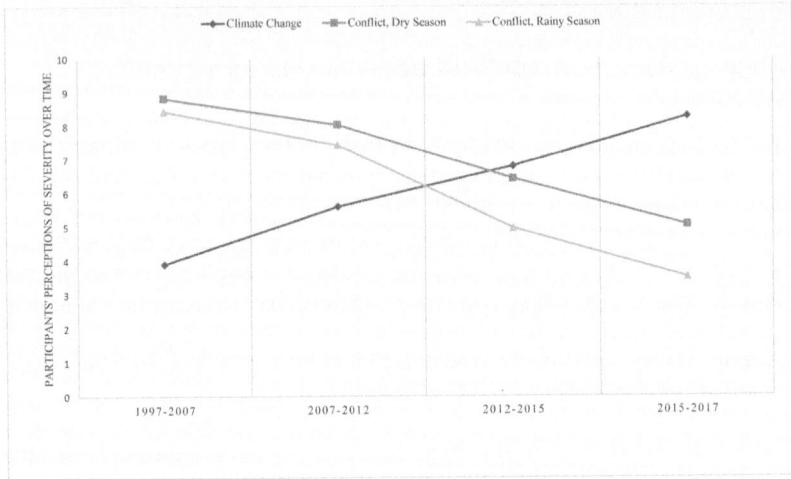

Source: author's survey questions focused on biophysical changes in Karamoja, n=103.[2]

Figure 6.1 Perceptions of severity over time

For example, in a typical example, an LC1 from Amudat described how changing precipitation patterns led to conflict outcomes:

> The rains are sometimes too much and sometimes too little. Sometimes it is delayed. It [the rainfall unpredictability] has led to large migration of animals in search of water and pasture and it has also led to serve hunger due to food shortage ... This has led to conflicts between the Pokot and Karamojong.

This pattern, whereby climate change is negatively affecting conflict in present-day Karamoja, is reflective of a number of seemingly contradictory dynamics. First, though climate change does appear to be having an impact on contemporary conflict, the severity of conflict has decreased over the past decade. Second, though ongoing regional changes related to conflict and development are generally presented in a positive manner, particularly by governmental and non-governmental agencies, they also have consequences. For example, following the region's stabilization, Karamoja's accessibility has rapidly expanded through an increasing network of paved roads. This accessibility has contributed to the increased rates of land grabs and resource extraction, key factors exacerbating vulnerability to climate change in Karamoja and in turn the higher rates of localized conflicts (Howe et al., 2015; Levine, 2010;

Pecquet, 2014). That is, and as I describe in detail below, the unique challenges of post-conflict recovery have, for some, led to increased vulnerability to environmental change and particular conflicts that threaten the region's stability.

Livelihoods, agency, and identity

Beyond just threatening the region's stability and contributing to the localized forms of conflict, the impacts of climate change, in conjunction with a wider set of political ecological factors, have disrupted livelihood systems in Karamoja. Indeed, 69 percent of research participants described the primary socioeconomic effects of climate change as negatively affecting agriculture; 57 percent described negative impacts on pastoralism, and 100 percent of participants described negative impacts on livelihoods as a result of climate change. How those outcomes manifested resembles how Nixon reconceptualizes 'displacement' as something more than being forced to move from one's place of belonging (Nixon, 2011, p. 19):

> [T]he loss of the land and resources beneath them [is] a loss that leaves communities stranded in a place stripped of the very characteristics that made it inhabitable ... Such a threat entails being simultaneously immobilized and moved out of one's living knowledge as one's place loses its life-sustaining features. What does it mean for people declared disposable by some "new" economy to find themselves existing out of place in place as, against the odds, they seek to slow the ecological assaults on inhabitable possibility? What does it mean for subsistence communities to discover they are goners with nowhere to go, that their once-sustaining landscapes have been gutted of their capacity to sustain by an externalizing, instrumental logic?

Reflecting Nixon's argument regarding displacement, changes in Karamoja have led to shifts in identity and sense of agency. As a result of economic shifts including a central-government-led push from pastoralism to agriculture, the impacts of disarmament, and the biophysical impacts of climate change, populations that have traditionally relied on pastoralism have been thrust into a socioeconomic system that no longer values their skills as highly. Many survey respondents described a shift in the physical and economic landscape following disarmament that seemed to not simply ignore but actively inhibit those that had traditionally been pastoralists. The attritional and deep-reaching slow violence of these shifts is epitomized in the experience of an LC1 from Iriri subcounty; his description was typical of LC1s in the region. In response to a question about the impacts of climate change, he noted:

> Pastoralism will be much affected in Karamoja and animals will reduce. Yet this is an industry which is a backbone for Karamoja livelihood and every other system will be disorganized.

He further explained how prolonged dry spells have diminished areas that used to be communal water reserves and that low agricultural productivity has led many people both out of and into Iriri: youth move to urban areas due to limited economic opportunities locally yet due to its relative high rainfall as compared to other areas, pastoralists come to Iriri in search of 'greener places.' Many of those who do not migrate face hardships and competition and usually resort to cutting trees for charcoal, which he believed is hurting soil degradation and driving erosion furthering this cycle. Like this respondent, participants from throughout the region described variations of how reinforcing and cascading set of vulnerabilities drove shifts in livelihoods, calling into question perceptions of one's agency and identity. For example, research participants regularly described increasing reliance upon what were once emergency sources of income, as primary sources (e.g., artisanal mining and charcoal production) due to unpredictable rainfall and declining viability of agriculture (see also Nakalembe et al., 2017). In explaining the extremity of these changes, an LC1 from Katikekile subcounty described the localized conflict over the mining sites:

> Peoples' livelihood has changed from agricultural products to natural resource products like gold, marble stones which was not the case before. Now people are struggling for sites where to mine from.

In Karamoja, livelihood systems, familial structures, and gender-oriented identities have shifted in complex ways. Though the ongoing shifts in the region, particularly those related to pastoralism, most directly affected men, the impacts are not confined to those who would traditionally be engaged in pastoralism. These shifts reflect that livelihoods are not simply products of economics, but decisions and outcomes born of context-specific factors related to agency, social structures, and economic motivations (Carr, 2013). For instance, Mercy Corps staff would occasionally point out the example of men brewing beer, traditionally a job performed by women. They did this both to show that men were increasingly giving up on pastoralism and demonstrating how women were being displaced from what was once an important source of income. Research participants, such as an LC3 from Napak district, reflected similar points, describing how changes in seasonal rainfall patterns have reduced the productivity and even the viability of pastoralism and agriculture leading to increasing reliance on new, and largely undesirable, ways to make a living.

> Rains used to experience in March and September but these days they are in April. There is too much sunshine these days. Like the sun has lowered down a bit. [Adding] This has made people fail to distinguish the correct planting time ... This

has led to burning of food in the gardens, people have abandoned farming and resorted to charcoal burning for survival.

As was consistently described by survey respondents, children and women had to deal with the cascading impacts of climate change on socioeconomic systems. Research participants described the impacts of climate change as leading to alcoholism, child labor, domestic violence, and sexual abuse. 25 percent of participants explicitly described alcoholism and household violence as an impact of climate change. In describing some of the more draconian impacts of how individual vulnerabilities and the impacts of climate change impact violence in particular populations, participants described how the decreasing viability of pastoralism led to high rates of alcoholism and domestic violence. Others described climate change as a factor leading to child endangerment. Though clearly not linked to climate change alone, these descriptions help contextualize how climate change can alter social structures, power dynamics, and agency with severe consequence for highly vulnerable populations, even if those populations are not directly affected by climate change. For example, an LC1 from Nadugnet described the impacts of climate change as follows:

> Due to failed agriculture, children are exposed to hard core activities like burning charcoal, carrying firewood to the market. Some families send their children to Moroto town to beg so that they can have something to eat.

Other participants described how a similar confluence of events that they linked to climate change were leading to increased rates of prostitution, early marriage, and sexually transmitted diseases. For example, when asked about impacts of climate change in his community an LC1 from Tapac subcounty noted:

> Many people have moved to Moroto towns to look for casual labor and many young children have been given for marriage.

Though I do not argue that climate change caused the outcomes described above, that participants linked the impacts of climate change to domestic violence and prostitution is telling. Whereas the violence of 'those days' closely resembled what Galtung (1969) described as direct violence, the post-conflict succession has led to new vulnerabilities that are insidious, cascading, difficult to comprehend, and call into question individuals' agency and identity. These challenges have led some residents to describe the peace in paradoxical terms: while their safety has improved, they have been left exposed to new risks that they are less prepared to face. Climate change, though exogenous, is embedded in those challenges.

Rhetorical comparisons

Returning to the question of how one can make slow violence visible, the data demonstrate that the rhetoric with which residents described present-day Karamoja is often contrasted, compared, or metaphorically linked with the violence of 'those days.' The armed conflict of 'those days' was severe. By some metrics, during the peak of this violence, Karamoja was one of the most violent places on the planet (Bevan, 2008). Research participants described harrowing examples of traumatic violence that they experienced personally. That includes their personal experiences in gun fights as 'warriors,'[3] being shot in armed ambushes, the need to run to shelter by a certain time to avoid the risk of stray bullets, and the devastating loss of family members. Likewise, the disarmament process that sought to quell this violence also led to deeply traumatic events, with allegations of rape, torture, and extrajudicial killings against the UPDF (Human Rights Watch, 2014; UNHCHR, 2007).

It is, therefore, striking that research participants regularly compared and contrasted the impacts of climate change and the unique post-conflict vulnerabilities to 'the gun' or to 'those days.' For example, when asked about how weather changes were being experienced in his village, an LC1 from Moroto District described how, in 'those days,' violent conflict was the biggest challenge to food security. He then described how, at present, extended and more intense dry seasons and unpredictable weather were his biggest challenges. This LC1 was happy to see conflict decreasing in severity, but was deeply concerned that new forms of vulnerability were manifesting, noting:

> It [used to be] only the gun which prevented people to cultivate, but now, the sun has become another gun.

Similarly, in reflecting on the economic progress in the region and how the region has changed in his lifetime, an LC1 was both dismissive of and frustrated with the changes being experienced in the region.

> People say there is development. I see the iron sheets. I see the tarmac. But I am pale. My muscles are thin. I don't take milk. Those days, even when there was the gun, we were under trees and I looked healthy. I was strong. We were taking milk. We were taking meat. I had cattle; I was a rich man. Now we are vulnerable.

These descriptions speak to region wide trends. Vulnerabilities exist in spite of, or perhaps even because of, the stabilization and associated economic changes in the region. These new vulnerabilities present a paradox: though many people feel safer, they feel more vulnerable due to insidious and new challenges that are harder to see and difficult to articulate. Drawing upon ontologies of slow violence provides unique insight into these latent and less apparent vulnerabilities.

Conclusion

In describing the ways climate change manifests as a form of attritional and insidious violence in Karamoja, this chapter builds on the body of work demonstrating the limitations of focusing on whether or not climate change alters conflict outcomes. That is, insomuch as climate change may or may not be a 'threat multiplier' in Karamoja, its less dramatic impacts on the quotidian are of considerable consequence even if easily overlooked. Research participants' consistent comparison of the contemporary vulnerabilities that have led to localized conflict outcome to 'the gun' or 'those days' is reflective of the severity and seriousness of slow violence.

Clear, too, is that just as direct violence rarely has just one cause, it is always multiple factors that lead to slow violence. In this way, it is revealing that research participants regularly described how the impacts of climate change not only affected those most proximate to the increasing variability (e.g., pastoralists and would-be pastoralists) but those most vulnerable in the region (e.g., young girls). This reflects that even in geographies as localized as a district or village, the effects of the slow violence of environmental change on a particular population will be far from uniform and will always interact with a wider set of determinants (Carr & Onzere, 2018).

In conceptualizing climate change as a stressor of biophysical and socio-economic systems the utility of the slow violence framework is clear. As is evident in this case, calling direct attention to the repercussions that are often disconnected in time and space, slow violence provides a focusing lens on the occluded relationship between the drivers and impacts of environmental change. In so doing, slow violence directly aligns with recent calls in political ecology and STS literatures for the deconstruction of the temporal, spatial, and discursive disconnections that mask the multi-scaled inequities of global climate change (Barnett, 2019; Le Billon & Duffy, 2018; O'Lear, 2016). This does, however, represent methodological challenges as to how one can see and measure the slow violence of climate change (or environmental change more generally). This research offers an example of how one can approach addressing this methodological challenge. An emphasis on descriptive data, utilizing discourse and rhetoric to identify metaphorical descriptions of slow violence, and triangulating data across multiple sources all were critical in making more apparent the unseen violence imparted by the effects of climate change.

Acknowledgments

I appreciate the assistance of Agnes Lotukei with the implementation of the survey in Karamoja. I also thank Kabir Singh for his assistance in coding and analyzing the data referenced in this chapter. I am grateful for the survey respondents and other research participants for their time and thoughtful responses. I acknowledge the support of Mercy Corps, in particular the Environment, Energy, and Climate, PEACE III, and BRACED teams who helped facilitate this research. The data collected for this chapter were supported by the United States Agency for International Development, Grant No. AID-OAA-G-16-00026, as part of a separate project. I also acknowledge the support of the Buck Lab for Climate and Environment at Colby College, which supported analysis costs and provided research assistance.

Notes

1. Daniel Abrahams was not at USAID when the research for the current paper was conducted. The views and opinions expressed in this paper are those of the authors and not necessarily the views and opinions of the United States Agency for International Development.
2. I consulted with Mercy Corps staff to identify benchmarked timeframes and used staggered timeframes to account for the challenges of human memory. The purpose, primarily, was to better understand how participants saw these changes over time.
3. Warrior was often used as a catch-all term for an armed pastoralist who took part in cattle raids.

References

Abrahams, D. (2018). *From Rhetoric To Response: Climate Change, Conflict, And Development In Karamoja, Uganda.* University of South Carolina. Retrieved from https://scholarcommons.sc.edu/etd/4990

Abrahams, D. (2020). Conflict in abundance and peacebuilding in scarcity: Challenges and opportunities in addressing climate change and conflict. *World Development, 132,* 104998. https://doi.org/10.1016/j.worlddev.2020.104998

Abrahams, D., & Carr, E. R. (2017). Understanding the connections between climate change and conflict: contributions from geography and political ecology. *Current Climate Change Reports, 3*(4), 233–242. https://doi.org/10.1007/s40641-017-0080-z

Barnett, Jon (2001). *The Meaning of Environmental Security: Ecological Politics and Policy in the New Security Era*. London: Zed Books.

Barnett, J. (2019). Global environmental change I: climate resilient peace? *Progress in Human Geography, 43*(5), 927–936. https://doi.org/10.1177/0309132518798077

Barnett, J., & Adger, W. N. (2007). Climate change, human security and violent conflict. *Political Geography, 26*(6), 639–655. https://doi.org/10.1016/j.polgeo.2007.03.003

Bevan, James (2008). *Crisis in Karamoja Armed Violence and the Failure of Disarmament in Uganda's Most Deprived Region* (No. 21). Geneva, Switzerland: Small Arms Survey.

Branch, A. (2018). From disaster to devastation: drought as war in northern Uganda. *Disasters, 42*, S306–S327. https://doi.org/10.1111/disa.12303

Bushby, Kirstin., & Stites, Elizabeth (2016). Resilience and risk in pastoralist areas: recent trends in diversified and alternative livelihoods, Karamoja, Uganda. In *Resilience and Risk in Pastoralist Areas: Recent Trends in Diversified and Alternative Livelihoods* (Peter Litt, pp. 11–31). Washington, DC: USAID Feed the Future.

Carr, E. R. (2013). Livelihoods as intimate government: reframing the logic of livelihoods for development. *Third World Quarterly*. https://doi.org/10.1080/01436597.2012.755012

Carr, E. R., & Onzere, S. N. (2018). Really effective (for 15% of the men): lessons in understanding and addressing user needs in climate services from Mali. *Climate Risk Management*. https://doi.org/10.1016/j.crm.2017.03.002

Chaplin, D., Byekwaso, F., Semambo, M., Mujuni, G., Bantaze, J., Nyasimi, M., … Krishnaswamy, S. (2017). *The Impacts of Climate Change on Food Security and Livelihoods in Karamoja*. Kampala, Uganda: UNMA, CGIAR, CCAFS, WFP.

Collier, P., Hoeffler, A., & Söderbom, M. (2008). Post-conflict risks. *Journal of Peace Research, 45*(4), 461–478. https://doi.org/10.1177/0022343308091356

Dalby, Simon (2009). *Security and Environmental Change*. Cambridge, MA: Polity Press.

Detges, A. (2014). Close-up on renewable resources and armed conflict. *Political Geography, 42*, 57–65. https://doi.org/10.1016/j.polgeo.2014.06.003

Eaton, D. (2007). The business of peace: raiding and peace work along the Kenya Uganda border (Part I). *African Affairs, 107*(426), 89–110. https://doi.org/10.1093/afraf/adm085

Eaton, D. (2008). The business of peace: raiding and peace work along the Kenya-Uganda border (Part II). *African Affairs, 107*(427), 243–259. https://doi.org/10.1093/afraf/adm086

Emerson, Robert. M., Fretz, Rachel. I., & Shaw, Linda L. (2011). Writing ethnographic field notes. In *Writing Ethnographic Fieldnotes* (2nd edn., pp. 21–41). Chicago and London: The University of Chicago Press.

Filipová, Z., & Johanisova, N. (2017). Changes in pastoralist commons management and their implications in Karamoja (Uganda). *Journal of Political Ecology, 24*(1), 881. https://doi.org/10.2458/v24i1.20972

Floyd, R. (2008). The environmental security debate and its significance for climate change. *The International Spectator, 43*(3), 51–65. https://doi.org/10.1080/03932720802280602

Galtung, J. (1969). Violence, peace, and peace research. *Journal of Peace Research, 6*(3), 167–191. https://doi.org/10.1080/10632921.2015.1103673

Gemenne, F., Barnett, J., Adger, W. N., & Dabelko, G. D. (2014). Climate and security: evidence, emerging risks, and a new agenda. *Climatic Change, 123*(1), 1–9. https://doi.org/10.1007/s10584-014-1074-7

Gray, S., Sundal, M., Wiebusch, B., Little, M. A., Leslie, P. W., & Pike, I. L. (2003). Cattle raiding, cultural survival, and adaptability of East African pastoralists. *Current Anthropology, 44*(S5), S3–S30. https://doi.org/10.1086/377669

Hecht, G. (2018). Interscalar vehicles for an African Anthropocene: on waste, temporality, and violence. *Cultural Anthropology, 33*(1), 109–141. https://doi.org/10.14506/ca33.1.05

Howe, Kimberley, Stites, Elizabeth, & Akabwai, Darlington (2015). *"We Now Have Relative Peace": Changing Conflict Dynamics in Northern Karamoja, Uganda.* Somerville, MA: Feinstein International Center.

Human Rights Watch (2007). *"Get the Gun!" Human Rights Violations by Uganda's National Army in Law Enforcement Operations in Karamoja Region.* New York: Human Rights Watch.

Human Rights Watch (2014). *"How Can We Survive Here?" The Impact of Mining on Human Rights in Karamoja, Uganda.* Washington, DC: Human Rights Watch.

Ide, T. (2017). Research methods for exploring the links between climate change and conflict. *Wiley Interdisciplinary Reviews: Climate Change, 8*(3), e456. https://doi.org/10.1002/wcc.456

Ide, T., Schilling, J., Link, J. J. S. A., Scheffran, J., Ngaruiya, G., & Weinzierl, T. (2014). On exposure, vulnerability and violence: spatial distribution of risk factors for climate change and violent conflict across Kenya and Uganda. *Political Geography, 43*(April), 68–81. https://doi.org/10.1016/j.polgeo.2014.10.007

Kaufman, A. (2014). Thinking beyond direct violence. *International Journal of Middle East Studies, 46*(2), 441–444. https://doi.org/10.1017/S0020743814000427

Le Billon, P. (2001). The political ecology of war: natural resources and armed conflicts. *Political Geography, 20*(5), 561–584. https://doi.org/10.1016/S0962-6298(01)00015-4

Le Billon, P., & Duffy, R. V. (2018). Conflict ecologies: connecting political ecology and peace and conflict studies. *Journal of Political Ecology, 25*(1), 239. https://doi.org/10.2458/v25i1.22704

Levine, Simon (2010). *What to do about Karamoja? A Food Security Analysis of Karamoja.* Rome, Italy.

Mach, K. J., Kraan, C. M., Adger, W. N., Buhaug, H., Burke, M., Fearon, J. D., ... von Uexkull, N. (2019). Climate as a risk factor for armed conflict. *Nature.* https://doi.org/10.1038/s41586-019-1300-6

Mamdani, M. (1982). Karamoja: colonial roots of famine in north-east Uganda. *Review of African Political Economy, 25*(Sep–Dec 1982), 66–73. https://doi.org/www.jstor.org/stable/3998097

Maxwell, Joseph (1992). Understanding and validity in qualitative research. *Harvard Educational Review, 62*(3), 279–301. https://doi.org/10.17763/haer.62.3.8323320856251826

Mercy Corps (2016). *Karamoja Strategic Resilience Assessment.* Portland, Oregon.

Mkutu, K. A. (2010). Complexities of livestock raiding in Karamoja. *Nomadic Peoples, 14*(2), 87–105. https://doi.org/10.3167/np.2010.140206

Mubiru, D. N. (2010). *The impacts of climate change on food security and livelihoods in Karamoja.* Rome, Italy: Food and Agriculture Organization of the United Naitons.

Nakalembe, C., Dempewolf, J., & Justice, C. (2017). Agricultural land use change in Karamoja Region, Uganda. *Land Use Policy, 62*, 2–12. https://doi.org/10.1016/j .landusepol.2016.11.029

Nakalembe, C., Dempewolf, J., and Justice, C. (2017). Agricultural land use change in Karamoja Region, Uganda. *Land Use Policy, 62*, 2–12. https://doi.org/10.1016/j .landusepol.2016.11.029

Nixon, R. (2011). *Slow Violence and the Environmentalism of the Poor*. Cambridge, MA and London, UK: Harvard University Press.

Oels, A. (2015). Resisting the climate security discourse. In S. O'Lear and S. Dalby (eds.), *Reframing Climate Change: Constructing Ecological Geopolitics* (pp. 188–202). Abingdon, UK and New York, NY, US: Routledge.

O'Keefe, M. (2010). Chronic crises in the arc of insecurity: a case study of Karamoja. *Third World Quarterly, 31*(8), 1271–1295. https://doi.org/10.1080/01436597.2010 .542968

Okpara, U. T., Stringer, L. C., & Dougill, A. J. (2016). Perspectives on contextual vulnerability in discourses of climate conflict. *Earth System Dynamics, 7*(1), 89–102. https://doi.org/10.5194/esd-7-89-2016

O'Lear, S. (2005). Resource concerns for territorial conflict. *GeoJournal, 64*(4), 297–306. https://doi.org/10.1007/s10708-005-5808-y

O'Lear, S. (2016). Climate science and slow violence: a view from political geography and STS on mobilizing technoscientific ontologies of climate change. *Political Geography, 52*, 4–13. https://doi.org/10.1016/j.polgeo.2015.01.004

O'Lear, Shannon and Simon Dalby (eds.) (2015). *Reframing Climate Change: Constructing Ecological Geopolitics*. London: Routledge. https://doi.org/10.4324/ 9781315759265

O'Loughlin, J., Witmer, F. D. W., Linke, A. M., Laing, A., Gettelman, A., & Dudhia, J. (2012). Climate variability and conflict risk in East Africa, 1990–2009. *Proceedings of the National Academy of Sciences of the United States of America*. https://doi.org/10 .1073/pnas.1205130109

Pecquet, J. (2014, February). Kerry doubles down on climate change. Secretary Kerry Warns Climate Change Is A "Weapon Of Mass Destruction." *The Hill*.

Peluso, N. L. (2008). A political ecology of violence and territory in West Kalimantan. In *Asia Pacific Viewpoint* (vol. 49, pp. 48–67). https://doi.org/10.1111/j.1467-8373 .2008.00360.x

Peluso, N. L., & Watts, M. (2001). *Violent Environments*. Ithaca, NY: Cornell University Press.

Ravallion, M. (2000). Carbon emissions and income inequality. *Oxford Economic Papers, 52*(4), 651–669. https://doi.org/10.1093/oep/52.4.651

Selby, J., & Hoffmann, C. (2014). Beyond scarcity: rethinking water, climate change and conflict in the Sudans. *Global Environmental Change, 29*, 360–370. https://doi.org/ 10.1016/j.gloenvcha.2014.01.008

Stark, Jeffrey (2011). Climate change and conflict in Uganda: the cattle corridor and Karamoja. *United States Agency for International Development*. Washington, DC: USAID.

Stites, E., & Akabwai, D. (2010). "We are now reduced to women": impacts of forced disarmament in Karamoja, Uganda. *Nomadic Peoples, 14*(2), 24–43. https://doi.org/ 10.3167/np.2010.140203

Turner, M. D. (2004). Political ecology and the moral dimensions of "resource conflicts": the case of farmer-herder conflicts in the Sahel. *Political Geography*. https:// doi.org/10.1016/j.polgeo.2004.05.009

UNHCHR (2007). *Report of the United Nations High Commissioner for Human Rights on the Situation of Human Rights in Uganda: Update Report on the Situation of Human Rights in Karamoja*. Geneva.

Verhoeven, H. (2011). Climate change, conflict and development in Sudan: global neo-Malthusian narratives and local power struggles. *Development and Change, 42*(3), 679–707. https://doi.org/10.1111/j.1467-7660.2011.01707.x

Wapner, P. (2014). Climate suffering. *Global Environmental Politics, 14*(2), 1–6. https://doi.org/10.1162/GLEP

Wapner, Paul, & Elver, Hilal (eds.) (2016). *Reimagining Climate Change*. Abingdon, UK: Routledge, Earthscan. https://doi.org/10.4324/9781315671468

Watts, Michael (1983). *Silent Violence: Food, Famine and Peasantry in Northern Nigeria*. Berkeley: University of California Press.

Watts, M. (2004). Resource curse? Governmentality, oil and power in the Niger Delta, Nigeria. *Geopolitics*. https://doi.org/10.1080/14650040412331307832

Willett, J. L. (2015). The slow violence of climate change in poor rural Kenyan communities: "Water is life. Water is everything." *Contemporary Rural Social Work, 7*(1), 39–55.

7   Enduring infrastructure

Kimberley Anh Thomas

Introduction

All along the lower reaches of the Mekong River, it was clear in late 2019 that something was afoot. People were paying more attention than usual to the long belt of water as it swept past the Laotian capital of Vientiane, curved along the northeastern border of Thailand, and merged with the Tonle Sap River in Cambodia. Tourists and fishermen alike were captivated by the river's sparkling, aquamarine waters. Sightseers in northeastern Thailand ventured out into the river to play on bare sandbanks, while laborers in central Laos drove trucks down to the water's edge and collected exposed rocks and sand for construction. It was a marvel to witness, yet it was all wrong. Riverside dwellers had never experienced anything like it, water so clear it reflected the color of the sky. The river, ever opaque with a muddy brown hue, was suddenly flowing as if filtered from a city tap. Despite the river's unexpected beauty, people's wonder quickly gave way to concern, and the hyaline waters became increasingly viewed with suspicion and fear.

It did not take long to attribute blame for the river's unprecedented appearance, as scientists and lay people quickly linked the clear water to the recently commissioned Xayaburi Dam in Laos. Under normal conditions, the spidery network of streams and rivers of the Mekong Basin discharges approximately 160 million tons of sediments into the East Sea every year. According to regional hydrologists and ecologists, the newly minted dam had trapped so much sediment that it starved the river's lower mainstem within a matter of weeks. The water became crystalline for lack of its usual cargo of silt.

Dams are notorious for displacing people and destroying habitats during dam construction and reservoir filling, as well as for altering downstream flows and impeding fish migration. However, less known are the additional and extensive risks that sustained sediment deprivation poses to riparian communities, both human and non-human. Notable among these is the 'hungry water effect,' whereby rivers robbed of sediment become hyper-charged with kinetic energy. These faster flowing currents scour riverbeds, erode banks, and destabilize vegetation and structures along the rivers' edge.

Fish and invertebrates that rely on sediments for habitat and nutrients are additional casualties of sediments held captive behind dams. The Mekong River ranks only second to the Amazon in terms of biodiversity, and these organisms sustain the largest inland fishery in the world. The Mekong supplies an astonishing 15% of the global freshwater catch, which provisions roughly 60 million fishing households with livelihoods and food security. Sediments of various sizes create microhabitats for a wide array of fauna to rest, lay eggs, and hide from predators, all essential for healthy aquatic systems. Yet, these critical ecosystem processes hinge on the delivery of water and sediment according to the steady tempo of the monsoon climate. Therefore, while anti-dam protests highlight the deleterious impacts of disrupted flows of water and fish, changes to long-standing patterns of sediment transport are equally significant for the structure and function of many riparian socioecologies.

In Southeast Asia, as elsewhere, hydropower dams generate electricity that courses along transmission lines into homes, schools, businesses, and hospitals. They impound water that may be conducted along irrigation canals and pipes and into rice fields and kitchen sinks. Dams may double as roads across deep valleys that may otherwise be impassable, thereby connecting towns and villages to each other, as well as to cities beyond. Such energy, water, and transportation infrastructures "comprise the architecture for circulation, literally providing the undergirding of modern societies, and they generate the ambient environment of everyday life" (Larkin 2013: 328). Yet, they are also critical sources and sites of violence.

The carnage of flattened homes, razed trees, clogged waterways, and skies choked with dust index the violence wrought by bulldozers and saws, wrecking balls and eviction notices. Researchers, journalists, and members of impacted communities have filled volumes documenting the spectacular offenses wrought by large infrastructure projects, particularly during and immediately after their construction. Substantially less attention has been paid, however, to the quiet violence that unfolds after infrastructure 'goes live.' Such violence operates over longer stretches of time and is less overt in its effects. As I signaled in my sketch of the lower Mekong River in the wake of the Xayaburi dam, my goal is to focus on these slower forms of violence inflicted upon complex socionatures through the quotidian operation of water management infrastructure.

Such inquiry poses a unique challenge, as slow violence by definition does not accommodate itself well to examination. By virtue of its temporality and invisibility it often goes undetected by all but its victims (Davies 2019). Even then, causal threads can be difficult to disentangle. In the process of articulating the insidiousness of slow violence, Rob Nixon (2011: 14) lays out the task through the language of apprehension:

> How do we bring home—and bring emotionally to life—threats that take time to wreak their havoc, threats that never materialize in one spectacular, explosive, cinematic scene? ... To engage slow violence is to confront layered predicaments of apprehension: to apprehend—to arrest, or at least mitigate—often imperceptible threats requires rendering them apprehensible to the senses through the work of scientific and imaginative testimony.

I approach this task from the presupposition that social and environmental systems are co-constitutive. It makes little sense in any case to attempt to cordon off 'nature' as somehow distinct from human spheres of life. This is especially true when considering the dialectical relationship between biophysical processes and social [re]production, as it is mediated by various infrastructures.

In the case above, non-human natures create the preconditions for human flourishing in the Mekong River Basin, supplying the fertile soils, diverse plant and animal life, water, climate, and abundant fisheries that have sustained human life for millennia. As part of complex socioecological systems, plant cultivation, fishing, harvesting, resource management, and other social practices have likewise shaped regional climates, resource availability, and the spatial and temporal distribution of various life forms. The introduction of extensive or large-scale infrastructures significantly alters these dynamics by redirecting material flows, often with violent social and environmental consequences. To 'apprehend' these consequences requires us to consider infrastructure in relational terms. In other words, infrastructures are not standalone objects but rather "articulated components" (Star and Ruhleder 1996: 111) within a complex web of relations.

Infrastructure studies have flourished since the mid-1990s, and we now have at our disposal a rich body of research interrogating the political and social lives of infrastructure. This work has rightly emphasized the assaults on human and non-human natures that occur during the course of infrastructural construction, abandonment, and decay. However, my goal here is to illuminate the violences enacted elsewhere during infrastructural life—namely, that beguiling period when infrastructures are humming along as planned, when the indispensable services they deliver obscure the socioecological costs of provision.

Indeed, it is difficult to conceive of everyday life without the systems that deliver drinking water to our homes, enable long-distance communication, and convey us between domestic, work, and recreational spaces. It is by virtue of the fact that these amenities have become essential to daily life that the full complement of their socioecological costs is neither captured in even the heftiest price tags of infrastructure projects nor widely appreciated by the diverse publics that rely on networked services. My argument here is that beyond the more visible and well-documented violence associated with infrastructural construction, abandonment, and deterioration, infrastructure also metes out gradual, accretive violence in the course of its normal, even optimal, functioning. As I illustrate through several examples of water infrastructure, violence is not incidental to infrastructure but intrinsic to it. This is lamentably the case because of the way that spatial, temporal, and social horizons are written into infrastructural designs. However, these same factors reveal opportunities for producing more egalitarian ecologies, thus pointing a way toward non-violent infrastructural relations. In the remainder of the chapter, I summarize relevant insights from ethnographies of infrastructure and interpret additional cases of infrastructural violence. I conclude by articulating key questions to guide investigations of slow violence in the aftermath of novel landscape engineering.

Social science of infrastructure

Anthropologists have advanced our collective understanding of infrastructure by theorizing the materiality, aesthetics, imaginaries, governance, and temporalities of infrastructure (see Larkin 2013; Anand et al. 2018; and Hetherington 2019 for helpful syntheses). According to this and related work in geography, infrastructure is less defined by its *thingness* than by its processes and relations (Cousins 2019). Even such unambiguously material objects as cables, satellites, aqueducts, roads, communication towers, and rail lines are embedded in networks of relations that regulate every aspect of their operation, ranging from their access and use to their functioning and upkeep. Infrastructures also require continuous inputs of labor, be they for planning, construction, maintenance, monitoring, or repair. In summary, they are brought into existence and sustained through a combination of environment-derived materials, human ingenuity, and labor.

To approach infrastructures relationally is to recognize that they are in a constant state of becoming. This ongoing-ness belies any assumed teleology in which infrastructure neatly progresses from conception to construction and culminates in completion or decommission. Indeed, the reality of infrastructural life is marked by fits and starts, hiccups, delays, and even abandonment. "If infrastructure is processual ... then one might conclude that finishedness is illusory—that everything is unfinished" (Carse and Kneas 2019: 13).

While this may be the case, many infrastructural projects realize some period during which they fulfill their intended functions. We do experience lights coming on, water flowing from taps, and trains running along tracks, after all. It is when systems operate as we expect them to that they have the tendency to slip into the background and become invisible, even when their enabling infrastructures—irrigation canals, telephone wires, pipelines—are plain for anyone to see (Harvey 2018). Thus, it is often, though not uniformly (Larkin 2013; Schwenkel 2015), observed that infrastructure is most visible when it malfunctions (Star and Ruhleder 1996, Edwards et al. 2009). The challenge of apprehension is doubled, therefore, by the difficulty of perceiving slow violence *and* the infrastructural media that engenders such violence. How then do we render visible the slow violence of infrastructure?

Star and Ruhleder emphasized infrastructure as a relational property when they posed the provocative question, "when is an infrastructure?" (1996: 112). I suggest that one way to apprehend its slow-onset harms is by reworking their query and asking: When is infrastructural violence? Or, as Rodgers and O'Neill (2012: 402) put it, "A key conceptual challenge, then, is to understand when it is that infrastructure becomes violent, for whom, under what conditions, and why." That infrastructural violence is enacted during construction, neglect, breakdown, and decay is well established. However, the creeping violence that unfolds when infrastructure 'works' is arguably more insidious and as such has escaped equivalent critical attention. I will argue that such violence is written into infrastructural blueprints due to the spatial, temporal, and social horizons around which infrastructures are designed. But first, I will present a few cases to illustrate how slow violence operates through infrastructure.

Infrastructural violence

A moment's consideration of the role of infrastructure in our lives quickly leads to an awareness of its ubiquity. We need look no farther than our own homes to recognize that infrastructure is integral to the distribution of water, electricity, information, people, manufactured goods, finance, food, and more. Among these, water is especially well-suited to elucidating the relationship between infrastructure and slow violence, as water itself is a "total social fact"—a social phenomenon that cuts across virtually all domains of society (Orlove and Caton 2010). Every aspect of life is touched in some way by water, and rarely does it arrive in our lives unmediated by a pipe, canal, sewer, pump, hose, or faucet. Despite our utter dependence on such infrastructure, "structural forms of violence often flow through material infrastructural forms" (Rodgers and O'Neill 2012: 405), and how these infrastructural forms articulate into sites and moments of violence varies over time. A good deal of research has focused on abuses arising at the inception and demise of infrastructural life, two moments onto which I map three categories of infrastructural violence: displacement, failure, and decay.

During the building of large-scale infrastructure, communities are often forced from their homes and the productive resources that support their livelihoods. Attending such displacement, communities may be relocated onto marginal land, un- or under-compensated for their losses, or subject to social and cultural disruption through relocation across dispersed sites. Such was the case for an ethnic minority of Laos, 2,700 of whom were resettled to clear space for hydropower development but without compensation and onto land already occupied by another minority group (Green and Baird 2016). Government officials typically rationalize such violence by invoking imperatives for development and progress, or increasingly climate change adaptation, but the damages are inflicted all the same.

A second set of grievances is associated with infrastructural failure. On one level, problems arise when projects are abandoned as a result of projects running out of funding or political winds shifting in favor of some alternative. Even incomplete projects may exact a toll, however, as communities suffer the effects of displacement, as well as landscapes scarred by habitat destruction and construction debris. Partial but non-functional structures can also persist as monuments of unfulfilled promises and serve as painful reminders of what could have been. Even when infrastructures do come online, they may fall woefully short of their intended service delivery in terms of expected quality or duration. One paradigmatic example is the catastrophic failure of Louisiana's levees during Hurricane Katrina in 2005 that resulted in over 1,500 deaths as well as untold social traumas that persist today.

Periods of neglect or decay mark a third notable moment of infrastructural violence. While technologists and capitalists champion innovation and disruption, others argue that the "mundane labor" of maintenance is what really warrants acclaim (Russel and Vinsel 2016). Maintenance keeps things running long after the sparkle of a novel infrastructure has worn off and people have come to expect and rely on timely performance. Maintenance also has indirect benefits, such as the social cohesion fostered among farmers in Egypt through their practice of communal upkeep of irrigation ditches (Barnes 2016). However, when neglected, as opposed to intentionally decommissioned, water infrastructure wreaks its own havoc beyond the termination of essential services. In Cambodia, for instance, combined sewer systems quickly become clogged with trash when pipes are not regularly cleaned. Congested pipes become overwhelmed during heavy rains, thereby increasing the occurrence of flooding and sewage overflows, as well as their attendant health risks (Jensen 2017). Human health and safety were even more dramatically sacrificed when up to 12,000 children in Flint, Michigan were exposed to lead leached from aging pipes. An entire generation of the predominantly African American population was subjected to gradual poisoning as a result of progressive devaluation and underinvestment in the city and its infrastructure (Ranganathan 2016).

This cursory survey merely hints at innumerable other cases of violence that unfold during various stages of infrastructural life. But if a key challenge to addressing slow violence entails rendering it visible, then it behooves us to focus on that period when infrastructure is most obscured from view: those times when it functions 'properly,' when it delivers services so fundamental to everyday activities that it is taken for granted as part of the background of social life. What violences do we begin to notice when we focus on infrastructure at its most benign?

The [un]working of water infrastructure

The most familiar water infrastructures are likely those systems that underpin domestic water supply and sewerage. However, as the example of levees and floodwalls in Louisiana indicate, people may be equally dependent on those landscape modifications that offer protection *from* water. Given that water is necessary for all life, it is unsurprising that 70–80% of the human population lives within 5 km of a waterbody (Kummu et al. 2011). This close proximity to water presents opportunities for health and sanitation, food production, manufacturing, and commerce, but it also poses greater risk. Striking the balance between optimizing the benefits of water while minimizing its hazardous potential has increasingly meant managing the timing, distribution, and quality of water using engineered solutions such as wastewater treatment, irrigation canals, pumps, and dams. Flood management structures in Vietnam, for example, powerfully convey the violence of deflected risk, but to understand this infrastructural violence requires first placing it within historical context.

Upstream dikes, downstream risk

Vietnam is the world's third largest exporter of rice, but this has not always been the case. The country periodically faced acute food shortages during the 20th century and was a net importer of rice as recently as 1985. Much of this dramatic shift in food security can be accounted for by looking to the Mekong Delta, where 90% of Vietnam's rice exports are sourced. Here, rich deposits of alluvial soil are highly conducive to agriculture, but their benefits were limited by the six-month monsoon season, during which the region receives over 90% of its annual rainfall and surface water, and severe floods can reach depths up to 3 m. In addition to contending with extreme seasonality that constrained what they could plant and when, farmers were also strongly influenced by political economic factors. Rice yields were particularly low under state-led attempts at collectivization of agricultural and industrial production, first starting in the North in the 1960s and then in the South after reunification in 1975 (Raymond 2008). Against a backdrop of chronic grain shortfalls and pervasive malnutrition, the central government implemented sweeping economic reforms in the late 1980s, including land and water management practices geared toward optimizing rice production.

This national 'rice first' food security policy entailed the widespread adoption of high-yield rice varieties and agrochemicals in conjunction with physical engineering of the hydrological regime to regulate fresh and salt water flows tuned to these modern rice varieties (Tran and Kajisa 2006). Water management schemes were modeled on Dutch polders, swaths of land encircled by earthen embankments for flood protection. Such structures had been introduced to the Mekong Delta in the 1930s, but the central government was not able to develop them in earnest until the economic reforms of the late 1980s (Biggs et al. 2009). The drive to increase rice production for domestic and international markets encountered significant challenges in the upper delta, where naturally deep floodwaters were not conducive to high-yield rice varieties that had short stems compared to the 'floating rice' typical of the region. Engineers and planners therefore implemented a vast network of embankments across the upper delta to exclude floodwaters from the paddy fields.

The dikes were immediately effective and instantly popular. Farmers could now plant multiple crops per year, boosting rice production 39%, from 3.03 tons per hectare in 1985 to 4.2 tons per hectare in 2000 (Le Coq and Trebuil

2005). Gains for farmers were short-lived, however, even though Vietnam has consistently generated an annual grain surplus of an average 3 million tons of rice. Trading companies have captured the lion's share of the benefits of market liberalization, while smallholder farmers have become one of the poorest segments of the population (Biggs et al. 2009). Part of the explanation for this disparity lies in the fact that in addition to requirements for chemical inputs associated with Green Revolution crops, the altered hydrology introduced its own costs. Namely, the installation of high dikes in the upper delta, particularly after disastrous floods in 2000, had the unfortunate but predictable consequence of excluding river-borne sediments from depositing on the land. As a result, farmers had to apply ever greater amounts of expensive fertilizers to compensate for the associated loss of nutrients. Moreover, flooded fields are an important fishing site for poor farmers and landless people, and the removal of the flood period has eliminated opportunities to save money from fishing, as well as time for farmers to rest while their fields sat fallow (Kakonen 2008). Accordingly, one recent study concluded that the shift to triple-cropping is only optimal for wealthy, large landowners and only for a period of about ten years, given the cost of replacing sediment-bound nutrients with artificial fertilizers as soil fertility progressively declines (Chapman and Darby 2016).

The gradual impoverishment of smallholders and soil fertility represents one type of slow violence brought on by the normal functioning of high dikes in the upper delta. However, the characteristic feature of water as a flow resource means that river systems are highly interconnected, and waterbodies rarely align themselves with the administrative units that people impose on them. For the central delta, this connectivity has translated into increased flood risk, as the onrush of monsoon water has to go somewhere. When high dikes in the upper delta confine floodwaters to the channels, the water will empty out at the first opportunity. Such has been the case for Can Tho, the largest city in the Mekong Delta, which experienced a 27% increase in annual water levels after a wave of dike construction in 2007 (Dang et al. 2016). In 2019, Can Tho experienced its worst flooding in 30 years as monsoon rains coincided with high tides, even though the city is 80 km from the coast. Roads and homes were inundated with muddy water, creating an array of hardships for 30,000 people who lost work, missed school, struggled to move around the city, and faced days of clean-up after the waters receded. High dikes have been instrumental to Vietnam's economic growth and food security in recent decades, but flood protection and rice production in the upper delta have come at the expense of rising inequality among farmers and greater flood risk and damages in unprotected areas downstream.

I will pause here for a moment to point out two salient dimensions of infrastructural violence based on the cases above. First, it has a *temporality*, which logically derives from the focus on slow violence. It takes time for soils to become depleted and for flood channels to clog with silt. It likewise takes time for agrarian livelihoods to become increasingly tenuous as the margin between agricultural inputs and rice yields narrows. Second, the displacement of flood hazards in the Mekong Delta reveals that infrastructural violence has a *spatiality*. An infrastructural system and the violent effects it engenders are not necessarily co-located. Regulating water in one area may cause spillover effects elsewhere. Rendering the slow violence of infrastructure visible thus entails adopting a wide spatial and temporal scope. Turning our attention to sea wall protection in Nigeria allows us to add to this list a third dimension of *sociality*—a who to the when and where of infrastructural violence.

Sacrifice zones

In many cases, the victims of infrastructural violence are readily identifiable. It is evident that poor and marginalized communities in New Orleans, children in Flint, and ethnic minorities in Southeast Asia have been crushed under the weight of failed levees, corroded pipes, and massive dams. But how do we account for those not easily recognized as casualties of infrastructural violence? By following the water, it becomes possible to see how the strict designation of a target population for service provision can increase the precarity of those located just beyond it.

The most populous and fastest growing city in Africa is pushing the boundaries of its growth. Like many coastal cities, Lagos, Nigeria only has so much room to expand. This geographical constraint helps to explain a massive land reclamation effort to develop a 10 km^2 residential and business complex adjacent to the affluent neighborhood of Victoria Island. It would seem that such an endeavor is well timed given the astounding housing deficit that has left 70% of the city's 21 million residents living in slums precariously located in wetlands and floodplains. Eko Atlantic's high-rise apartment buildings will accommodate up to 300,000 people once complete, but the luxury units will do nothing to alleviate the affordable housing crisis. Instead, the developers of the US$6 billion undertaking have their sights set on those in an entirely different income bracket. The exclusive enclave is billed as "one of the world's cutting-edge new cities" and is modeled on Manhattan's skyscraper district (ekoatlantic.com). It also boasts the Great Wall of Lagos, an 8.5 km-long sea revetment designed to protect the artificial peninsula from flooding, sea level rise, and coastal erosion.

Flood and erosion protection are noteworthy features given that Eko Atlantic was first conceived in 2003 as a retaining wall to help Lagos as a whole contend with these very hazards (Brisman et al. 2018). Plans for collective shoreline protection were later reconfigured into a narrowly focused sea defense barrier that only shelters the new city and a portion of Victoria Island. Constructed of 100,000 five-ton concrete blocks, the 25-ft-high sea wall is intended to withstand 1 in 1,000-year storm events, as its design also took "into consideration global warming and rising sea levels" (Eko Atlantic 2017). However, beyond the wall's covered range, unprotected and predominantly slum areas face greater threats from erosion, sea level rise, and coastal flooding, as the sea wall deflects incoming waves and storm surge down the shore (Ajibade 2017).

As in the Mekong Delta, sea water carried shoreward by storms and tides will find a path to expend itself. The difference here is one of stark social differentiation, whereby environmental protection for the wealthiest comes at the expense of heightened risk for impoverished groups who already suffer higher mortality from recurring and intensifying floods (Thomas and Warner 2019). "Even more than other gated communities, Eko Atlantic says to the world, *No, we are not all in this together*" (Goodell 2017: 219). Of course, it is possible to rank the risks of flood damage and erosion based on property values, according to which prioritizing Eko Atlantic for sea wall protection makes economic sense. Indeed, such a pointed calculus is what keeps Lagos off the list of top megacities most at risk from sea level rise, in spite of the sizable population there likely to be displaced by inundation, as other cities like Guangzhou, Mumbai, and New York have more valuable infrastructure (Goodell 2017). However, cost-benefit analysis also leads to outcomes like the well-being of hundreds of thousands of residents in floating slums being sacrificed to enhance the security of those already well-poised to weather Nigeria's more frequent storms. The ubiquity of this approach in infrastructural planning means that there are many more examples of such cold calculation and harsh outcomes elsewhere in the world, and they are likely to proliferate, particularly where social inequity is high.

The importance of social difference to slow violence is underscored in a recent article that challenges Rob Nixon's (2011) assertion that slow violence persists due to a paucity of compelling stories. Thom Davies (2019: 13) mobilizes ethnographic work on toxic landscapes in Louisiana to argue instead that "slow violence persists because those 'arresting stories' do not count. Crucially, a politics of indifference about the suffering of marginalized groups helps to sustain environmental injustice, allowing local claims of toxic harm to be silenced." Building on his conclusions, we can recognize that the accounting behind the design of the Great Wall of Lagos facilitates infrastructural violence through two distinct mechanisms. It discounts the suffering of the city's most vulnerable from the outset to justify callous decisions in terms of rational economics, and it ensures that any telling of residents' ensuing struggles do not count. The fact that the communities most impacted by the sea wall are the city's poorest is of central importance, as their marginal political status affords them no recourse against the development's powerful stakeholders.

Scripted violence

In her clear-eyed examination of the ongoing violence of water poisoning in Flint, Malini Ranganathan (2016) invites her audience to read against the grain of environmental racism. Her motivation for doing so is that, while environmental racism is frequently invoked as an explanation for the water crisis, such an interpretation encounters two pitfalls. One either runs into a burden of proof problem when attributing environmental racism to individual prejudice, or one takes race for granted and neglects to address the causal forces behind its production. Instead, she lays bare liberalism's entrenched contradictions and ambivalences to demonstrate how "racial hierarchy is foundational—and not simply incidental—to the workings of capitalism and an ostensibly democratic, liberal market society" (Ranganathan 2016: 5).

In the cases described above, I sacrificed depth for breadth in the hope of staking a parallel claim about infrastructural violence. Slow violence is not an exception to the smooth operation of infrastructure, an unfortunate by-product of otherwise well-conceived plans, but is constitutive of it. I submit that slow violence is hardwired into infrastructure. It is written directly into every infrastructural blueprint, mock-up, model, and promo video. And this is where the three dimensions of infrastructural violence come into play.

From the outset, every project envisions and demarcates a social, spatial, and temporal scope: a population of beneficiaries, a target region, a time horizon for service delivery. This is standard practice for any infrastructural design. However, it is also a techno-managerial mechanism by which slow violence becomes normalized, because what may lie beyond any of these social, spatial, and temporal horizons is of little to no consequence to the project. All those spillover effects and unintended outcomes that manifest as slow violence either do not register or are diminished in the logic of generating essential services. Shifts in sediment loads, soil nutrient concentrations, stream velocity, sea level, and water chemistry may be imperceptible to our blunt senses, but they are cumulative and mightily consequential to those who suffer their terrible effects.

They can also be surprisingly predictable if you know where and when to look. Few today are surprised that dams decimate fish populations, jetties wash away beaches, and dikes redistribute flood risk, for example. Yet, these interventions continue to be implemented all over the world under the banner of modernization, climate change adaptation, and progress. That they are perpetually introduced as solutions despite their repeated failings demands greater scrutiny of the logics that underpin them (Thomas 2020). Peering into the social, spatial, and temporal workings of infrastructure thus becomes a necessary act of revolt: "To imagine and to make environments of justice, then, is necessarily to engage in the 'boring' technopolitics of infrastructure; to reveal, refuse, and revolt against the ways in which their vital and violent politics are frequently obscured and buried from view" (Anand 2017). Envisioning socially and environmentally just alternatives to violent infrastructures entails broadening the social, spatial, and temporal horizons that overly circumscribe who is to benefit from or pay the socioecological costs of infrastructural services, where, and for how long.

Accounting for infrastructural violence

This chapter advances two key arguments. First, infrastructure has long been recognized to be a key site and moment of violence, and, increasingly, slow violence. However, this awareness tends to revolve around infrastructure at its inception and demise, particularly during episodes of displacement, failure, and decay. Attention to the period when infrastructure functions as intended and provides valuable services reveals another, yet underexamined, array of violent effects. Second, infrastructures give rise to slow violence through the delineation of time scales, spatial areas, and social groups that are strictly focused on a network's immediate outcomes, service delivery region, and customers. Lag times and knock-on effects for human and non-human communities outside these boundaries are woefully omitted from infrastructural design processes, and thus the creeping injustices they suffer remain unaccounted.

While it is likely unfeasible to account for every instance of infrastructural violence, many of its dynamics have been well characterized, and it may therefore be possible to anticipate and intercept slow violence in the making. Interrogating planners and proponents about a project's social, spatial, and temporal scope is one important starting point. What is the full spectrum of downstream impacts of the infrastructure? How will the system disrupt existing socioecological dynamics, and with what effects? Who is responsible for mitigating violence once it comes to light? To whom is the project responsible? What would make it responsive to non-beneficiaries in non-target areas? Such questions anticipate slow violence from the outset and help stretch an infrastructure's social, spatial, and temporal horizons to encompass all those who pay the dreadful costs of their vital services.

Acknowledgments

I wish to thank Jenny Goldstein for taking the time to read and provide thoughtful comments on a draft of this chapter.

References

Ajibade, I. (2017), 'Can a future city enhance urban resilience and sustainability? A political ecology analysis of Eko Atlantic city, Nigeria', *International Journal of Disaster Risk Reduction*, **26**, 85–92.

Anand, N. (2017), 'The banality of infrastructure', *Items*, 27 June, accessed 10 July 2017 at https://items.ssrc.org/just-environments/the-banality-of-infrastructure/

Anand, Nikhil, Akhil Gupta, and Hannah Appel (eds.) (2018), *The Promise of Infrastructure*, Durham, NC: Duke University Press.

Barnes, J. (2016), 'States of maintenance: Power, politics, and Egypt's irrigation infrastructure', *Environment and Planning D: Society and Space*, **35**(1), 146–164.

Biggs, D., F. Miller, C. T. Hoanh, and F. Molle (2009), 'The Delta machine: Water management in the Vietnamese Mekong Delta in historical and contemporary perspectives', in F. Molle, T. Foran, and M. Kakonen (eds.), *Contested Waterscapes in the Mekong Region Hydropower, Livelihoods and Governance*, London: Earthscan, pp. 203–225.

Brisman, Avi, Nigel South, and Reece Walters (2018), 'Climate apartheid and environmental refugees', in Kerry Carrington, Russel Hogg, John Scott, and Maximo Sozzo (eds.), *The Palgrave Handbook of Criminology and the Global South* (2nd ed.), Cham, Switzerland: Springer International Publishing, pp. 301–321.

Carse, A. and D. Kneas (2019), 'Unbuilt and unfinished', *Environment and Society*, **10**(1), 9–28.

Chapman, A. and S. Darby (2016), 'Evaluating sustainable adaptation strategies for vulnerable mega-deltas using system dynamics modelling: Rice agriculture in the Mekong Delta's An Giang Province, Vietnam', *Science of the Total Environment*, **559**(C), 326–338.

Cousins, J. J. (2019), 'Malleable infrastructures: Crisis and the engineering of political ecologies in Southern California', *Environment and Planning E: Nature and Space*, **0**(0), 1–23. http://doi.org/10.1177/2514848619893208

Dang, T. D., T. A. Cochrane, M. E. Arias, P. D. T. Van, and T. T. de Vries (2016), 'Hydrological alterations from water infrastructure development in the Mekong floodplains', *Hydrological Processes*, **30**(21), 3824–3838.

Davies, T. (2019), 'Slow violence and toxic geographies: "Out of sight" to whom?', *Environment and Planning C: Politics and Space*, **4**(3), 239965441984106–19.

Edwards, P. N., G. C. Bowker, S. J. Jackson, and R. Williams (2009), 'Introduction: An agenda for infrastructure studies', *Journal of the Association for Information Systems*, **10**(5), 364–374.

Eko Atlantic (2017), 'Eko Atlantic's "Great Wall of Lagos" passes 6km', accessed 17 January 2019 at https://www.ekoatlantic.com/latestnews/press-releases/eko -atlantics-great-wall-lagos-passes-6km/

Goodell, J. (2017), *The Water Will Come: Rising Seas, Sinking Cities, and the Remaking of the Civilized World*, New York: Little, Brown and Company.

Green, W. N. and I. G. Baird (2016), 'Capitalizing on compensation: Hydropower resettlement and the commodification and decommodification of nature–society relations in Southern Laos', *Annals of the American Association of Geographers*, **106**(4), 853–873. http://doi.org/10.1080/24694452.2016.1146570

Harvey, P. (2018), 'Infrastructures in and out of time: The promise of roads in contemporary Peru', in Nikhil Anand, Akhil Gupta, and Hannah Appel (eds.), *The Promise of Infrastructure*, Durham, NC: Duke University Press.

Hetherington, K. (ed.) (2019), *Infrastructure, Life, and Environment in the Anthropocene*, Durham, NC: Duke University Press.

Jensen, C. B. (2017), 'Pipe dreams: Sewage infrastructure and activity trails in Phnom Penh', *Ethnos*, **82**(4), 627–647.

Kakonen, M. (2008), 'Mekong Delta at the crossroads: More control or adaptation?', *Ambio*, **37**(3), 205–212.

Kummu, M., H. de Moel, P. J. Ward, and O. Varis (2011), 'How close do we live to water? A global analysis of population distance to freshwater bodies', *PLoS ONE*, **6**(6), e20578–13.

Larkin, B. (2013), 'The politics and poetics of infrastructure', *Annual Review of Anthropology*, **42**(1), 327–343.

Le Coq, J.-F. and G. Trebuil (2005), 'Impact of economic liberalization on rice intensification, agricultural diversification, and rural livelihoods in the Mekong Delta, Vietnam', *Southeast Asian Studies*, **42**(4), 519–547.

Nixon, R. (2011), *Slow Violence and the Environmentalism of the Poor*, Cambridge, MA: Harvard University Press.

Orlove, B. and S. C. Caton (2010), 'Water sustainability: Anthropological approaches and prospects', *Annual Review of Anthropology*, **39**(1), 401–415.

Ranganathan, M. (2016), 'Thinking with Flint: Racial liberalism and the roots of an American water tragedy', *Capitalism Nature Socialism*, **27**(3), 17–33.

Raymond, C. (2008), '"No responsibility and no rice": The rise and fall of agricultural collectivization in Vietnam', *Agricultural History*, **82**(1), 43–61. http://doi.org/10 .3098/ah.2008.82.1.43

Rodgers, D. and B. O'Neill (2012), 'Infrastructural violence: Introduction to the special issue', *Ethnography*, **13**(4), 401–412.

Russel, A. and L. Vinsel (2016), 'Hail the maintainers', *Aeon*, accessed 27 April 2020 at https://aeon.co/essays/innovation-is-overvalued-maintenance-often-matters-more

Schwenkel, C. (2015), 'Spectacular infrastructure and its breakdown in socialist Vietnam', *American Ethnologist*, **42**(3), 520–534.

Star, S. L. and K. Ruhleder (1996), 'Steps toward an ecology of infrastructure: Design and access for large information spaces', *Information Systems Research a Journal of the Institute of Management Sciences*, 7(1), 111–134.

Thomas, K. A. (2020), 'The problem with solutions: Development failures in Bangladesh and the interests they obscure', *Annals of the American Association of Geographers*, **0**(0), 1–21. https://doi.org/10.1080/24694452.2019.1707641

Thomas, K. A. and B. P. Warner (2019), 'Weaponizing vulnerability to climate change', *Global Environmental Change*, **57**, 101928.

Tran, T. U. and K. Kajisa (2006), 'The impact of Green Revolution on rice production in Vietnam', *The Developing Economies*, **44**(2), 167–189.

8 Slow violence and its multiple implications for children

Sheridan Bartlett

Introduction

While many chapters in this book tease out the concept of slow violence, expanding our understanding of the term and its nuances, I will settle here for taking the term at face value, borrowing Rob Nixon's definition before I move on to consider what this means for children. Nixon speaks of slow violence as "neither spectacular nor instantaneous, but rather incremental and accretive, its calamitous repercussions playing out across a range of temporal scales" (Nixon 2011, p. 2). He refers primarily to such gradually unfolding environmental debacles as climate change, deforestation, polluted atmospheres, and acidifying oceans. But he recognizes that the same qualities of gradual and often normalized harm pertain also to the kinds of structural violence that are built into our social systems and that manifest themselves through various kinds of exclusion and inequity that can sabotage people's capacity to realize their hopes and their potential. Fundamental to his account of slow violence is an appreciation of the porous boundaries between the metaphor of structural violence and the more immediate personal harm that we generally associate with the term violence. The stressful conditions generated by the former can in turn contribute to the latter.

I mention "moving on" to consider the implications for children, a nod perhaps to the dimension of time that is so central here and to the different time scales involved. Children "move on" so very quickly. They grow and develop in a fraction of the comparatively extended stretches of time implied by the slow violence in question. These rapidly developing children can be caught in the mill of slowly progressing catastrophes which, with our prime-time mentalities, can be perceived almost as non-events. A significant factor here is the progressively more chronic nature of what might in the past have been more episodic. Hundred-year floods are increasingly annual floods and annual floods can stretch and morph into persistent calamities. Conflicts can persist for many years. Displacement, whether on account of climate extremes or conflict, rather than being a relatively short-term event, can be the only reality a child has ever known.

We are accustomed to thinking of the disconnect between the time frame of slow violence and that of our political cycles, a practical reality that often derails even the most rational hardheaded proposals. We do not as often call to mind the disconnect between slow moving disruptions and the trajectory of children's developmental potential. There are plenty of more immediate problems to concern us for one thing. When events move faster than families and other social systems can adapt to, it is well established that children often face the most acute impacts. But when detrimental changes happen slowly, or linger, or weave themselves into our social systems, they can be just as lethal. Families and communities may be better equipped to accommodate and adapt to slow violence than to more extreme disasters, but too often the accommodation in question is only a matter of adjusting to worsening conditions, not actually tackling them. Either way, this accommodation can have especially dire consequences for children. This is not a matter of a few percentage points here and there. Several years ago, for instance, WHO estimated that 85 percent of all deaths attributed to climate change are among children (Mathers et al. 2009).

Children's disproportionate vulnerability

There are reasons why young children especially are so disproportionately at risk – why in the face of either slow violence *or* more rapidly unfolding events, they are more likely to die, to fall ill, to be injured, to endure emotional harm. Young children have underdeveloped immune systems, immature organs, more rapid metabolisms, more surface area relative to their mass, more limited experience and understanding. All these things leave them less well-equipped to deal with the varieties of stress and deprivation than is generally true for their elders.

Given their stage of rapid growth and development, there can also be serious long-term implications. Children's exposure to many of the manifestations of slow violence can have physical and psychological effects that play out over years. This is true regardless of whether these manifestations are the hardships of climate change, the toxic exposures related to pollution, or the insufficiencies that result from poverty, exclusion, and discrimination. Where an adult can recover from a season of food scarcity, for instance, a three- or four-year-old is more critically affected. The undernutrition that results may be accompanied by stunted growth, impaired health, compromised cognition, all with lifelong implications. Young children do not always just bounce back, and the likelihood of poor developmental outcomes increases exponentially with an accumulation of risk factors (Evans et al. 2013). Older children may be less emphatically affected, but they also suffer particular impacts, especially in terms of the life opportunities they are offered or can take advantage of.

Things have been getting better for the world's children on many fronts. Infant mortality, malnutrition, child labor, access to education, levels of poverty have all shown marked improvement over recent decades in most of the world (Boerma et al. 2018; UNICEF 2019). But as the impact of climate change has become gradually more extreme, as the associated conflicts and migration become more prevalent, and as disparities continue to widen, there have also been disheartening indications that in some places these trends are leveling out or even following a declining path. Consider, for instance, that since 2012, the number of girls and boys engaged in unacceptable forms of labor has increased by 10 percent. (FAO UN 2018). The global under-five mortality rate is a noteworthy bellwether. It has declined dramatically in recent decades – in 1990 there were 93 deaths for every 1,000 live births; by 2017, this was down to 39 deaths per 1,000 (UN Interagency Group for Child Mortality Estimation 2018). Yet national-level figures can mask the situation in smaller geographical areas or among particular groups, and in some places, especially in parts of Africa, these rates are stagnating or even climbing again (Li et al. 2019).

Impacts for health and survival

Malnutrition is one of the culprits. After a decade of steady improvement, it is on the rise again, primarily because of the increase in climate-induced stress and the conflicts that are so often related to this (FAO et al. 2017). In Asia, this is a matter of the rate of improvement slowing down. In South America and much of Africa, the situation is actually getting worse. Nearly one child in four is now chronically malnourished and 52 million suffer from acute malnutrition (Meybeck et al. 2018).

Globally, the trend without climate change is towards steadily decreasing rates of undernutrition for children under five, as is true for mortality. If climate change is factored into the equation, using the A2 scenario, projections indicate that an additional 25.2 million children under five will be affected by 2050 (Phalkey et al. 2015). This does not fully erase the gains that would otherwise have been made, but rather than a 30 percent decline in the number of undernourished children, the decline is projected to be less than 7 percent. Simon Lloyd and colleagues find that the most significant impact by 2050 will be for rates of severe stunting, which they estimate will increase by 23 percent in Central Africa, and by a massive 62 percent in South Asia (Lloyd et al. 2011).

Malnutrition, as just noted, has far-reaching implications. It is not simply a matter of catching up when supplies of food improve. Not only are stunted and wasted children more vulnerable to other causes of ill-health, they are also less likely to survive the added stress of an extreme event. Chronic malnourishment is also associated with effects for cognitive capacity, for school achievement, and for long-term earning power (Engle and Fernandez 2010).

Malnutrition is not solely a function of access to food. It is related also to unsanitary conditions, which can undermine a child's capacity to make good use of the food that is available. This is because of the resulting uptick in diarrheal disease and because of the energy drain for a child in fighting off infection – both of these squander calories which might otherwise go to growth and development. There is an interesting example from Mali, where there has been a general improvement in agricultural production over recent decades, as in much of the Sahel. Yet child mortality in Mali actually increased between 2000 and 2010 – a trend that researchers describe as "both puzzling and troubling" (Han and Foltz 2013, p. 3). It appears that an increase in rainfall in Mali, in part responsible for the improved food production, has also played a part in disrupting precarious sanitation arrangements. Where sanitation infrastructure is inadequate, an intense rainfall, more and more the norm in many places, can mean overflowing latrines, clogged storm drains, and contaminated water supplies. The incidence of diarrheal disease goes way up in these situations and young children are the most vulnerable.

A bout of diarrhea does not sound that ominous, but diarrheal disease actually remains one of the leading killers of young children in poverty and, like malnourishment, it can sabotage their more general health and development. Despite impressive reductions in diarrhea-related deaths since 2000, children younger than five years of age remain about twice as likely as to die from this common problem as the population at large, and they are also disproportionately vulnerable to all the associated morbidity (Keddy 2018).

Childhood malnutrition also remains the leading risk factor for mortality from lower respiratory infections for children under five and was estimated to be responsible for over 60 percent of these deaths in 2016 (Troeger et al. 2018). Children's vulnerability to these respiratory infections is related to a range of other factors too, including proximity to traffic, indoor crowding, and the cooking or heating fuel in use. It is also directly related to ambient outdoor air quality. The younger children are, the more serious is the impact of polluted air, because neither their lungs nor their immune systems are sufficiently developed to cope. The Global Burden of Disease figures show the death rate from lower respiratory disease for children under five to be more than three times higher than for the population at large (Troeger et al. 2018). Even children in the womb are at risk from polluted air, with higher rates of stillbirth, preterm birth and low birthweight. And this affects their development once they are born. Prenatal growth has been associated, for instance, with children's cognitive performance when they reach school age (Walker et al. 2007).

Air pollution is an excellent example of the tight relationship between the slow violence inherent in environmental deterioration and problems of social exclusion and inequity. In Ulaan Bator, Mongolia, one of the most polluted cities in the world, pollution levels are almost 30 times higher than the levels that WHO deems to be safe, and can at times climb even higher than that. This pollution was scarcely noticeable 20 years ago. Now the density of the smog makes midday as dark as evening. The pollution stems primarily from the expanding informal settlements surrounding the city. Almost half of Mongolia's population lives here now, a rural exodus driven by increasingly severe winters and drier summers, the loss of crops and animals, and the need for alternative livelihoods. These settlements are unserved by the centralized heating and other amenities that are present in the formal inner parts of the city. Almost two thirds of the city's more marginalized residents live in these peripheral areas, where they burn raw coal for heat and cooking. When coal runs short, as is frequently the case for these often-impoverished households, they burn plastic bottles and rubber tires to cope with the sub-zero temperatures. Year round, the air is not as bad as that in Delhi or Beijing, but in winter pollution levels climb to dangerous heights.

The situation has created a child health crisis. Fetal death rates point to the intensity of the threat – they are 3.5 times higher in winter than in summer. Children experience especially high rates of pneumonia, bronchitis, and asthma, and it is not unusual for them to miss school or to spend winters in and out of hospital. Overall, there has been an almost threefold increase in respiratory ailments over the last ten years, and this increase is highest for children, whose lung function was found to be 40 percent lower than that of children in rural areas. Long-term consequences are projected to include chronic respiratory conditions, reduced lung function, school failure and diminished productivity (Gheorghe et al. 2018).

Another threat is the spread of vector-borne diseases as a result of gradually warming conditions. The Centers for Disease Control and Prevention report a tripling of these diseases since 2004 in the USA alone. Despite heartening gains in fighting malaria over the last decade, it remains the most serious vector-borne disease globally in terms of mortality, and here again young children represent the great majority of victims (WHO 2017). Mortality rates alone, however, fail to capture the larger significance of the disease for this age group. An analysis of 48 African demographic surveillance studies early in the 2000s found that, because malaria results in chronic anemia, it also increases the severity of other diseases, more than doubling overall death rates for young children in high prevalence areas (Snow et al 2004). In addition to affecting overall health and growth, malaria can have other developmental implications. Results of a recent controlled study looking at the impact of malaria at age two for children's development at age five found a significant effect for the ability to cope with a range of cognitive tasks, as well as children's fine motor ability and socio-emotional skills (Fink et al. 2013).

Impacts for family stability

Although the direct impacts for young children are deep and troubling, an equally profound concern is the potential for these slow-moving events to deepen poverty, undermining families' capacity to get ahead, educate their children, improve their homes, consolidate their livelihoods, and climb out of poverty. There is growing evidence that the accumulation of everyday hardships is in fact taking a greater toll on household stability than that imposed by more dramatic and extreme events (Bull-Kamanga et al. 2003). When people are already struggling, every extended dry spell, every day spent coping with increasingly challenging conditions, can erode the gains that might have been made, destroying fragile assets. In the context of deteriorating physical conditions, what might otherwise be a manageable challenge can become truly overwhelming. The same is true for the range of structural pressures that can intensify exclusion, making day-to-day subsistence an increasingly tough proposition.

It is also significant that the forms slow violence can take are so often closely intertwined. The children and households and communities that experience one form of hardship are also more likely to be subjected to others. The urban households that most often suffer the impacts of climate change, for instance, are those living in informal settlements on flood plains or steep hillsides, built there because no other land is available, and in marginalized areas unserved by the drains and all-weather roads that protect their wealthier neighbors. The links between climate change and conflict are also illustrative. Conflicts are frequently generated or exacerbated by disputes over water and other resources, worsened by scant rainfall and high temperatures. Research in Africa found that a temperature increase of just one percent leads in that same year to a 4.5 percent increase in the likelihood of civil war, and to an additional 0.9 percent increase the following year. By 2030, say these researchers, drawing from data averaged across 18 climate models, this is projected to add up to a 54 percent increase in the incidence of armed conflict in the region (Burke et al. 2009).

The gradual erosion of the environments and circumstances that sustain the lives of children and their families can aggravate and intensify the many injustices that already burden the lives of the poor. And when households are under pressure, children are often the end point where the costs accumulate. One way of assessing these costs for children is through something as straightforward as counting calories. Research from Bangladesh looked at how food was distributed within households in poverty, first calculating the basic caloric needs of individual family members. When a household was managing reasonably well, any calories in excess of basic needs tended to be given to children. But when there were insufficient funds to purchase enough food to meet everyone's needs, adults were generally given priority and children received less than they needed to sustain their health and growth. In these situations, maintaining wage earners' capacity to keep on earning is of primary importance (Cockburn et al. 2009).

When families and communities are under great pressure, whether in the day-to-day struggle to stay ahead of changing climate conditions or in the context of chronic worry over the likelihood of eviction, depression and anxiety among adults becomes a serious concern (Patel and Kleinman 2003). Together with depleted resources, maternal stress can seriously affect the quality of care for young children. The sheer drudgery that accompanies inadequate housing and provision is at the best of times extreme – managing inadequate water supplies, dealing with waste and excreta, keeping children safe and clean in surroundings that are anything but that. Some element of "neglect" is almost inevitable in these conditions.

When I lived in Kathmandu early in the 2000s, I spent time visiting an illegal riverside settlement, where the constant threat of eviction, the lack of service provision, and rising river levels combined to create a precarious, taxing environment. Open ditches criss-crossed the settlement, carrying much of the city's liquid waste down to the garbage-filled river and often spilling over into the crowded settlement. It was almost impossible to keep small children out of the muck without confining them to their cramped shacks. If a ball bounced into a ditch, they often jumped right in after it, and their exhausted mothers, who started lining up for water from a distant standpipe at 3 a.m. each morning, tended to turn a blind eye. Diarrhea, rashes, and conjunctivitis were endemic. During the annual monsoon, the river rose a bit higher each year and flooded the shacks, filling them with a foot or more of foul-smelling mud. Families camped out for four or five weeks under sheets of plastic up on the nearby road until the rains stopped and the mud dried enough to be shoveled out. This community of 100-odd families became more impoverished every year. They could not approach the authorities for help; eviction would have been the most likely response. Neglect, I noted, can have an inevitable quality in the context of these burdens. But abusive treatment can also follow. Back in the 1990s, Richard Gelles found a clear connection between an accumulation of stressful events and circumstances in a household, and the use of violence against children (Gelles 1992). The capacity to manage stress is central here. Affluent families are certainly not immune to violence, but they can summon their resources to help manage pressure. A sense of control in life is basic to robust functioning, and for households dealing with slow violence, the capacity to manage stress can be in short supply.

It is not just the youngest children who suffer when family stability is challenged. Older children can be highly vulnerable to the strains, and many of them claim that it is the humiliation of their marginal status that is the hardest to accept. The longitudinal Young Lives study of children growing up in poverty in four countries has pointed repeatedly to the psychosocial impact for this age group, and to the sense of shame and stigma that can be so acute (Woodhead et al. 2013). In the Kathmandu settlement, older children would never reveal where they lived when they went to school – they found it too mortifying to identify themselves with this wretched scrap of land. Most of these children *did* at least go to a nearby school though. All too often, older children become assets that their families depend upon to maintain stability. As households become more constrained by circumstances, they are increasingly likely to turn to their children for work that can be damaging to their well-being and longer-term opportunities. In fact, another useful metric for assessing the strain that families are operating under, in addition to calorie counting, is whether, in the calculus of household stability, they are able to make a net investment in their children's future, or whether, on the contrary, they are forced to extract value from them. Child labor, which declined steadily for years, has, as noted, become more prevalent again especially in Africa, driven in part by an increase in protracted conflicts and by the climate-related stresses that often fuel them (FAO UN 2018). (It should be noted here that the term "child labor" does not refer to the more acceptable forms of employment or family-based work that are appropriate to a child's age and that do not challenge health or safety or deprive that child of schooling.) In countries affected by armed conflict, the incidence of child labor is estimated to be 77 percent higher than in the rest of the world, and especially hazardous child labor is 50 percent higher. In both categories, almost half of those involved are children between the ages of five and 11 (ILO 2017).

In many cases, children migrate for work unaccompanied by family. Shahin Yaqub's 2009 overview reveals how common a practice this is, and how very young the children can be. In Nepal, for instance, she found that 8 percent of all children between five and 14 became independent migrants for work; in Benin an astonishing 22 percent, averaging ten years of age. The sobering figures are repeated for country after country. In some cases, these children are helping their families; others may be seeking primarily to expand their own life chances in a context where opportunities are increasingly scarce at home; still others may be fleeing abuse in households where the stresses are extreme. In many cases children display surprising resilience, but the costs and the challenges can be extreme. Many end up on the street or engaged in prostitution or other damaging forms of survival (Yaqub 2009). These boys and girls are often discussed as victims of trafficking, and often this classification is indisputable. But whether they are unequivocally exploited or acting as purposeful agents, there is no question that they are victims of forces beyond their personal control.

Displacement

More and more, the chronic stresses of slow violence are culminating in dislocation for entire households as well. Forced global displacement has grown by almost a third over the last decade. In 2018, the number of displaced people in the world, whether refugees or those internally displaced in their own countries, rose to 70.8 million, the highest yet. Over half these people are children. About one child in 45 is estimated now to be on the move (UNHCR 2019). The ongoing debacle at the US border with Mexico is a well-documented reminder of the desperation and persistence of those escaping impossible conditions at home, and of the burden borne by children inexcusably separated from their parents.

Of course, many of the world's displaced people are fleeing from extreme events rather than the less acute manifestations of slow violence. Yet an important factor here, as noted before, is the progressively more chronic nature of these calamities, which might in the past have been brief or episodic. There is increasingly a blurring of the lines between extreme events and persistent or cumulative hardship. Neither disasters nor conflict nor life in emergency camps are properly what is meant by slow violence. And yet the very fact that one disaster runs into another, that conflicts, often driven by climate stress, become endemic, that emergency camps designed as short-term solutions become the only home that many children have ever known, pushes these issues into the realm of the chronic and the slow. Dabaab refugee camp in Kenya, for instance, which houses refugees from Somalia, has been open now for almost 30 years, and many people there have children and grandchildren who were born in the camp. Some arrived to escape civil war. Others arrived more recently when Somalia was hit by drought and famine. These camps, where residents are often prevented from taking employment outside, can be dysfunctional and even toxic environments where the notion of slow violence becomes especially poignant.

Not only is conflict, and the displacement that can accompany it, increasingly chronic. It also takes its toll more and more on civilians rather than combatants. Children make up the majority of casualties, not so much from weapons of warfare as from the deterioration in conditions that cause preventable diseases to flourish and families to become ever more stressed systems. However extreme the precipitating event may be, the long exhausting business of recovering a grip on life becomes for many households the more daunting experience. Mindy Fullilove, a psychiatrist who focuses on the mental health of communities, argues that displacement is the 21st century's defining problem. She refers to the associated suffering as "root shock" and describes this as "the traumatic stress reaction to the destruction of all or part of one's emotional ecosystem" (Fullilove 2005, p. 11) Her reference point here is the endlessly unfurling impact of urban renewal on the African American population in the USA, but her argument is relevant to millions globally.

Children as active agents

A final significant aspect of the relationship between children and slow violence is children's potential as active participants in the quest for social justice, and as responsible stewards of our challenged environments. There are compelling reasons for involving children in these efforts. They tend to have a lively awareness of their surroundings, and often bring a fresh perspective to local problems. The success worldwide of various well-managed participatory efforts with children around the ages of about eight to 15 has been amply documented.[1] After the 2004 tsumani, for instance, it was always children in the devastated Tamil Nadu communities I visited who pointed out that the rows and rows of replacement housing that were being built on the now-barren landscape would be uncomfortable with no shade, and that replanting trees should be a priority. Trees meant cool shade and fruit, but they were also where people gathered and where children played, an indispensable feature of local social and environmental capital. Adults and NGOs, meanwhile, focused on housing to the exclusion of all else, and no replanting had been planned.

There is a strong possibility, however, that there is a sensitive period as children age for the optimal development of indignation, commitment, and a sense of responsibility for the challenges we face. In the Philippines, ten-year-olds who lived in the grimy malodorous vicinity of the huge, notorious Payatas dump were outraged by the conditions they had to endure and wondered why the president had done nothing to help. Meanwhile, their older brothers and sisters had little to say about the situation, except to note their relief at having a reliable source of employment as rag pickers (Aguirre 2005). But even for children in this sensitive period, slowly deteriorating conditions can dull the awareness of change. Children especially, with their more limited time frame, may simply accept what are for them accustomed conditions. Peter Kahn referred to this limitation as "environmental generational amnesia" (Kahn 2002). Drawing from several studies, he noted how quickly people accommodate to worsening conditions, and how a sense of what is normal can change from generation to generation. Slow violence, then, may also play a role in undermining the development of the very awareness and commitment most needed in the creation of an engaged citizenry to tackle it.

Conclusion

In the Kathmandu settlement and in so many places around the world, it is the poorest families and communities that are most affected – the people who build on flood plains, who lack secure tenure, who have no storm drains, who cannot count on water supplies, whose crops fail again and again, who have the least capacity to prepare, adapt, and protect themselves or to invest in alternatives. The more gradual and persistent ordeals, whether expressed through repeated flooding or droughts or endemic conflict or the varied spatial and social expressions of inequity and exclusion, can mean rising prices for basic supplies, more precarious livelihoods, fewer opportunities, and more demands on time and energy to cope. This arrangement by which the world's most afflu-ent people transfer the cost of their lifestyles onto the nations and households that have done the least to contribute to changing climates is perhaps the most dramatic global manifestation of the slow violence inherent in inequity.

The research implications of the various impacts of slow violence for girls and boys are relatively simple. Just as gender has become an essential dimension to consider for a more nuanced understanding of social phenomena, so age also needs to be incorporated as a filter. An intersectional approach can involve, of course, a recognition of the particular implications for a whole range of special interests, all of them unquestionably important and deserving of attention. But it is important to bear in mind that children are not a minority group – in most of the world they make up over a third of the population. Their welfare is fundamental to society's longer-term health, and failing to take account of their presence and their particular requirements and potential can result in distorted findings and short-sighted policies. In the area of slow violence in particular, with its easily muted implications in the context of more rapidly moving world events, the longer-term consequences need close attention, and this involves attention to the determinants of children's current welfare as well as to the shape of the world they will inhabit.

Note

1.   See for instance the excellent collection of papers in four separate issues of the journal *Children, Youth and Environments* in 2006–2007.

References

Aguirre, Angela Desiree (2005), 'Local governance, children and the physical environ-ment: Payatas in the Philippines', *Children Youth and Environments* **15** (2), 138–150.
Boerma, T., J. Requejo, C.G. Victoria, and A. Amouzou (2018), 'Countdown to 2030: tracking progress towards universal coverage for reproductive, maternal, newborn and child health', *The Lancet* **392** (10152), 1072–1088.
Bull-Kamanga, L. et al. (2003), 'From everyday hazards to disasters: the accumulation of risk in urban areas', *Environment and Urbanization* **15** (1), 193–203.
Burke, Marshall B., Edward Miguel, Shanker Satyanath, John A. Dykema, and David B. Lobell (2009), 'Warming increases the risk of civil war in Africa', *Proceedings of the National Academy of Sciences* **106** (46), 10670–20674.

Cockburn, John, Anyck Dauphin and Mohammd A. Razzaque (2009), 'Child poverty and intra-household allocation', *Children Youth and Environments* **19** (2), 36–53.

Engle, P. L. and P. D. Fernandez (2010), 'INCAP studies of malnutrition and cognitive behavior', *Food and Nutrition Bulletin* **31** (1), 83–94.

Evans, Gary, Dongping Li and Sara Whipple (2013), 'Cumulative risk and child development', *Psychological Bulletin* **139** (6), 1342–1396.

FAO UN (2018, 12 June), 'Child labour in agriculture is on the rise, driven by conflict and disasters', accessed 16 September 2019 at http://www.fao.org/news/story/en/item/1140078/icode/

FAO, IFAD, UNICEF, WFP, and WHO (2017), *The State of Food Security and Nutrition in the World 2017: Building Resilience for Peace and Food Security*, Rome: Food and Agriculture Organization.

Fink, Günther, Analia Olgiati, Moonga Hawela, John M Miller and Beatrice Matafwali (2013), 'Association between early childhood exposure to malaria and children's pre-school development: evidence from the Zambia early childhood development project', *Malaria Journal* **12**, accessed 16 September 2019 at https://malariajournal.biomedcentral.com/articles/10.1186/1475-2875-12-12#auth-5

Fullilove, Mindy Thompson (2005), *Root Shock: How Tearing Up City Neighborhoods Hurts America, and What We Can Do About It*, New York: Ballantine Books.

Gelles, Richard (1992), 'Poverty and violence towards children', *American Behavioral Scientist* **35** (3), 258–272.

Gheorghe, Adrian, Batbayar Ankhbayar, Henlo van Nieuwenhuyzen, and Rogerio de Sa (2018), *Mongolia's Air Pollution Crisis: A Call to Action to Protect Children's Health*, Ulaan Bator: National Center for Public Health and UNICEF.

Han, Peter and Jeremy Foltz (2013), 'The impacts of climate shocks on child mortality in Mali', prepared for presentation at the Agricultural & Applied Economics Association's 2013 AAEA & CAES Joint Annual Meeting, Washington, DC, 4–6 August, 2013.

ILO (2017), *Global Estimates of Child Labour: Results and Trends, 2012–2016*, Geneva: ILO.

Kahn, Peter H. Jr. (2002), 'Children's affiliations with nature: structure, development and the problem of environmental generational amnesia', in Peter Kahn and Stephen Keller (eds.), *Children and Nature*, Cambridge, MA: The MIT Press, pp. 93–116.

Keddy, Karen H. (2018), 'Old and new challenges related to global burden of diarrhea', *The Lancet: Infectious Diseases* **18** (11), 1163–1164.

Li, Z. et al. (2019), 'Changes in the spatial distribution of the under-five mortality rate: small area analyses of 122 DHS surveys in 262 subregions of 35 countries in Africa', *PLoS ONE* **14** (1), e0210645.

Lloyd, Simon, Sari Kovats, and Zaid Chalabi (2011), 'Climate change, crop yields and undernutrition: development of a model to quantify the impact of climate scenarios for child undernutrition', *Environmental Health Perspectives* **119** (12), 1817–1823.

Mathers, Colin, Gretchen Stevens, and Maya Mascarenhas (2009), *Global Health Risks: Mortality and Burden of Disease Attributable to Selected Major Risks*, Geneva: World Health Organization.

Meybeck, A., E. Laval, R. Lévesque, and G. Parent (2018), 'Food security and nutrition in the age of climate change', *Proceedings of the International Symposium organized by the Government of Québec in collaboration with FAO, Québec City, 24–27 September 2017*, Rome: Food and Agriculture Organization.

Nixon, Rob (2011), *Slow Violence and the Environmentalism of the Poor*, Cambridge, MA: Harvard University Press.

Patel, Vikram and Arthur Kleinman (2003), 'Poverty and common mental disorders in developing countries', *Bulletin of the World Health Organization* **81**, 609–615.

Phalkey, Revati K., Clara Aranda-Jan, Sabrina Marx, Bernard Höfle, and Rainer Sauerborn (2015), 'Systematic review of current efforts to quantify the impacts of climate change on undernutrition', *PNAS* **112** (30), published online 27 July 2015 at https://www.pnas.org/content/early/2015/07/21/1409769112

Snow, R. W., E. L. Korenromp, and E. Gouws (2004), 'Pediatric mortality in Africa: *Plasmodium falciparum* malaria as a cause or risk?', *American Journal of Tropical Medicine and Hygiene* **7** (2) supplement, 68–75.

Troeger, C. et al. (2018), 'Estimates of the global, regional, and national morbidity, mortality, and aetiologies of lower respiratory infections in 195 countries, 1990–2016: a systematic analysis for the Global Burden of Disease Study 2016', *The Lancet: Infectious Diseases* **18** (11), 1191–1210.

UN Interagency Group for Child Mortality Estimation (2018), *Levels and Trends in Child Mortality Report 2018*, UNICEF, World Health Organization, World Bank Group, and United Nations Population Division.

UNHCR (2019), *Global Trends: Forced Displacement in 2018*, Geneva: United Nations High Commission for Refugees.

UNICEF (2019), The State of the World's Children 2019 Statistical Tables, accessed 8 November 2019 at data.unicef.org/resources/dataset/sown-2019-statistical-tables/

Walker, Susan P. et al. (2007), 'Association of growth in utero with cognitive function at age 6–8 years', *Early Human Development* **83**, 355–360.

WHO (2017), 'Vector-borne diseases', accessed 16 September 2019 at https://www.who .int/news-room/fact-sheets/detail/vector-borne-diseases

Woodhead, Martin, Paul Dornan, and Helen Murray (2013), *What Inequality Means for Children: Evidence from Young Lives*, Policy Paper, Oxford: Young Lives.

Yaqub, Shahin (2009), *Independent Child Migrants in Developing Countries: Unexplored Links in Migration and Development*, Innocenti Working Paper 2009–2001.

9

For Indigenous youth: towards caring and compassion, deconstructing the borderlands of reconciliation

Joseph P. Brewer II and Jay T. Johnson

Introduction

In the spirit of civil rights leader John Lewis, we are keeping his commitment to speak out, to "say something," when we see that things are not right. The lasting implications of colonial violence in and on Indigenous communities is slow, insidious, and a constant force that is both experientially tangible as well as intangible to those who are forced to engage with it on a daily basis. The central tenant of this chapter is a fundamental question we as authors have and continue to explore in our scholarship, that is how racist actions continually plague our global society in insidious and violent ways. For us, this question has implications in the study of, and really the manifestation of, the painful experience of violence. This is a kind of violence that for some really never goes away, you cannot hide from it, it is always in your face. While many scholars work to understand how scholarly questions have implications for the study of slow violence, the reality is that for populations whose ancestors experienced hate in the same way their relatives alive today experience it, the need to trudge through the weight of anguish that hate creates in order to strengthen the narrative of awareness is, for some, an unrelenting hope that change is on the horizon. Though, at this moment in history one may look across this nations landscape to see statues and memorials that celebrate the horrific treatment of African American slaves, or countless public lands named after individuals who worked to orchestrate the genocide of Indigenous peoples, such as Custer State Park. For those who live in these geographies and drive by these statues, memorials, and parks, they might be reminded on a daily basis of these events and histories, which for some is an intimate, recent history. How do they experience them? Is it that they just drive by without any feelings or thoughts of the significance of those places? When the lived histories of these places are told to their children, how do they experience the trauma as they drive by daily? It is a generational trauma that can be handed from grandparent to parent to child, lived out daily as a slow violence that most of society seems oblivious to, and/or chooses to ignore or not engage with. Is slow violence reconcilable? Can we as a society find common ground to transform it, and can we strengthen, restore, or create belief in an ideal of freedom and justice?

An extensive scholarship exists from Indigenous peoples and allies around the world about reconciliation and transformative and restorative justice. Much of this scholarship calls upon the settler-colonial state to acknowledge and commit to telling the truth about its history, and to move beyond the discomforts of truth-telling in order to initiate some action towards justice. Our recommendation for those who do not know that literature is to find it and read it, away from the disciplinary lens of academia, and more towards a lens of humanity. For those reading this piece, before assigning social labels such as "angry" or "dissident" because this chapter identifies social ills which can lead to discomfort, keep in mind that justice, in the context of human rights violations, for those who have suffered is often acknowledged when those who benefit from social injustice become aware. Let us pose two questions in the same spirit: What perpetuates acts of racism towards Indigenous people, regardless of age? How is it that future generations of Indigenous people come to know racism in similar ways that their great-grandparents knew it? Our answers are likely to reveal that reconciliation and transformative and restorative justice can only be realized when those who are the benefactors of others' pain can come to terms with their inheritance, speak truth to it, open themselves up to others' experiences, and try to heal. What is more often the case, however, is that when those in power are confronted with uncomfortable truths, they prefer to label those who help to strengthen the voice of the voiceless in ways that dehumanize those in the pursuit of justice.

There has been a call for and by Indigenous peoples, and many more ethnic and racially marginalized groups, to think about, express, and critique what justice is, has been, or can be for Indigenous people. As authors, we reflect on our life experiences and observations trying to understand the application of justice. We reflect on things that we had thought were past indiscretions, now looked back on as stepping stones, which have assisted our own intellectual growth, away from the struggles of adolescence and towards the sanctity of rational thought. We reflect on various learning opportunities. Reflect on the bumps and bruises that we all face and learn from if we are given time, the right skills, and access to a community that can assist with our reconciliation. Reflecting on these matters, this continues to be a long road for both of us. As Indigenous academics, our training tells us to hone in on, unpack, and deconstruct the discourses of reconciliation and transformative and restorative justice in order to identify the intellectual players of these concepts as they relate to slow violence. Inevitably, however, this only pulls us further and further away from the social and moral underpinnings of our foundation – the application of justice.

In order to talk about justice, in any form, in the U.S., we need to acknowledge, address, talk about, and create action-orientated plans to reconcile, transform, and restore integrity for instances of injustice. We need to ask ourselves why the (physical, spiritual, and emotional) harms Indigenous people endure/ endured at the hands of a long-lasting colonial agenda have not yet been addressed at the national level. We need to ask how Indigenous people will ever heal if their experiences are only recognized under a general consensus that Indigenous people have been treated "unfairly," with no plan to reconcile, transform, or restore what has been done.

This truth may be disheartening, but until mainstream U.S. society actively commits to learning and acknowledging the real unaltered truth about Indigenous people from Indigenous perspectives, the continued struggle and marginalization may fester. Efforts by those in power may continue to dehumanize those seeking justice. The truth is that, when ongoing murders and unsolved cases of missing and murdered Indigenous women as well as Indigenous lands leaving Indigenous hands continue, "sorry" isn't enough. Justice requires action. When referencing Indigenous communities in accompaniment with any form of justice, it is paramount to associate oneself with their unappropriated struggles.

Justice, appropriated and currently defined and represented by many settler-colonial ideologies in the modern world as a misguided superiority over all things, makes justice for Indigenous peoples a "hard sell." In fact, it is not a stretch to assert that justice for the settler-state, when judging matters directly or indirectly related to Indigenous people's rights, asserts itself above all other forms of justice, natural or imagined. As long as Indigenous people's human rights are subjected to a legal system that is largely uninformed, and Western society demonstrates a willingness to listen only to those who can shape their ethos in a palatable way for Western interests, justice will remain a struggle shaped by how comfortable advocates of ethnicity, race, socioeconomic class, and sexuality can make the settler-state when change is on the horizon. Real change or real justice, then, in those circumstances is erroneous.

A foregone conclusion wrapped in the discourses of politics, reliant on elected government officials and not on core values or natural laws of caring and compassion. Maybe, then, justice is better demonstrated or exemplified on a smaller social scale, away from the bright lights of politics and more towards the everyday interactions among members of communities. Here the hope and practice of justice for and by Indigenous people lies within foundational systems of reciprocity, focused on a more-than-human world. Within these interactions are the moral roots of justice, whether acknowledged or not, to leave this world in better shape than it was left to us, for future generations of human and non-humans alike. A task that is, at its core, about an investment in all life. Perhaps those at the national level interested in justice can be educated by examples from local communities to formulate protocols for justice nationally. Perhaps they can spend time learning about the racist behavior Indigenous children and communities are confronted with daily and help to do something meaningful and lasting. We are asking for inclusivity in the form of care and compassion to be of paramount priority for all manner of life.

What is transformative justice, exactly?

Transformative justice is about genuinely demonstrating actions of caring and compassion. This sense of caring and compassion though has been interrupted within a justice system that can seem to be more concerned with punishment than social justice. Within the settler-state justice system, caring for the physical, mental, emotional, and perhaps even the spiritual needs of the other has been fundamentally displaced away from, in essence, being human. So, what does transformative justice look like? In its truest form, it looks like you care, that you are compassionate enough to recognize, and take the time to do the hard work associated with, compassion.

For many, transformative justice is a way to provide an intellectual platform for the true intentions of restorative justice, to restore, repair, and reconcile with and for those who have been wronged (McAlinden 2011). The idea of *transforming* how or why justice is applied in criminal, civil, or regulatory law is important work, but not the kind of discussion we are interested in participating in here. Though scholars and advocates of justice are debating what transformative and restorative justice is and how it applies or more closely aligns with various sects of criminal justice (Marshall 1999), there are Indigenous communities who have been practicing caring and compassion as the foundation of restorative or transformative justice for millennia. For example, the Diné Peace Court system is a formal way to settle justice issues that arise in Diné communities, imploring Diné ideologies and epistemologies of *Hozho* to restoration. Peacemaking for Diné is about "The moral force of the group ... used to persuade people to put the group's good above individual welfare. It is said of a wrongdoer that 'he acts as if he had no relations'" (Yazzie and Zion 1996: 162). While this may look familiar to transformative and restorative justice practitioners and intellectuals, what is different is the deep spatial and temporal development of these intellectual traditions in practice. Often, the common experience for Indigenous communities that developed caring and compassionate ways to address circumstances that arose in their communities, as all societies have these experiences of disorder among community members, was to pursue ways to mediate while restoring meaningfulness to relationships communally. This approach is distinct from modern ideas and structures of restoration, but like most concepts used to describe what is naturally human, humans reshape, turn, deconstruct, and manipulate these to the point where we forget their original intentions. We do not want to do that here, so Indigenous examples are, in our opinion, more proven, mature methods of restorative justice for Indigenous people, and maybe for the world. It may also be, that these methods are flexible enough to incorporate other methods to help strengthen the overall goals of care and compassion. For, in reality, restorative justice is only a portion of justice in Indigenous communities (Gray and Lauderdale 2007).

There are many examples of Indigenous communities that formalized sophisticated protocols of restorative justice, such as the Haudenosaunee (Mohawk 1992, 2010), Anishinaabe (McGregor 2009, 2012), Lakota, Dakota, and Nakota (Deloria 1998) and Delaware (Weslager 1989). Justice, at least for Indigenous people, generally speaking, was accessible at all times, as Gray and Lauderdale (2007: 222) write: "For most North American Indians, law was accessible: the oral tradition allowed it to be carried around as part of them, rather than confined to legal institutions and inaccessible experts who largely control the language as well as the cost of using the law." Anybody interested in seeing how order plays out today in tribal communities need simply look at contemporary tribal codes, protocols, and processes that have been developed in the spirit of these practices (Brewer and Kronk Warner 2014). As Kyle Whyte et al. (2017: 183–191) have written, the fairly recent pursuit to seek Indigenous participation in conservation efforts needs to incorporate the "principle of self-determination; principle of early involvement; principle of intergenerational involvement; principle of continuous cross-cultural education; principle of balance of power and decision-making; principle of respect for Indigenous knowledge; and principle of control of knowledge mobilization." Whyte and his colleagues, through their extensive and ongoing research with tribal communities, have established these principles. A careful look at these principles, described also in the work of Leroy Little Bear (2000), clearly demonstrates these values are primarily born from a spiritual and religious construction that Indigenous people followed and still endeavor to follow. They are indeed from an orderly construction of a living universe, based within inclusivity and diversity. This "being together" with all of creation requires an understanding of the inherent interconnections upon which humans are dependent (Deloria and Wildcat 2001; Larsen and Johnson 2017). The Indigenous world is about order, and a world without communally structured order is out of balance, allowing chaos and movement away from natural practices of caring and compassion. So, in this chapter we are not interested in participating in the scholarly outcomes of defining or relating transformative or restorative justice to the ongoing rhetorical debates. If anything, we are interested in restoring or strengthening Indigenous youth's confidence in all of humanity, in a world that can seem to work towards inhumanity. While intellectualizing transformative and/or restorative justice allows for a clearer vision to interpret, hopefully understand, maybe empathize, and speak to slow violence, again the last thing we are trying to do is get pulled into the rhetorical semantics of defining slow violence absent of real-world applications. What we want, with our very being, centers on positive inclusive change, and if investing in the development of slow violence is a way to accomplish that, then here we are.

A beautiful genealogy

The Oceti Sakowin Oyate is a confederacy made up of three divisions: (1) the Dakota consisting of the Mdewakantonwan, Wahpekutetonwan, Wahpetonwan, and Sisitonwan families or Oyates, (2) the Nakota consisting of the Ihanktonwan and Ihanktonwanna Oyates, (3) the Lakota, also referred to as the Titonwan, consisting of the Oglala, Sicangu, Oohenonpa, Itazipco, Sihasapa, Hunkpapa, and Oyates. Today, the varying divisions of the Oceti Sakowin Oyate are primarily located on reservations in the states of South and North Dakota, Montana, Minnesota, and Canadian provinces. The Lakota youth, living on the Pine Ridge and Standing Rock reservations in South and North Dakota, are the focal point of this chapter.

The descendants of the Lakota divisions living on reservations throughout South and North Dakota have inherited an extraordinarily rich and beautiful culture that they strive to preserve and hand down to younger generations. Under unimaginable settler-colonial pressures intended to forcibly institute policies aimed at destroying Lakota identity and people all together, the Lakota are still here, alive, though not without their struggles, working towards self-determination and sovereignty. Some of the more damaging policies created by federal agencies and others of similar governmental pedigree since the U.S. inception include: off- and on-reservation boarding schools that gave federal officials the legal right to take children of various ages away from their parents for acculturation; the mass sterilization of Indigenous women; land tenure policies that made legal the theft and dissolution of millions of acres of land from Indigenous people; to name a few. Each of these examples has its own history, agents, and intention associated with it. Some policies and intentions are more divisive, some were based upon good intentions, and everything in-between. So, it may be true that when discussing Indigenous people's colonial histories in the U.S. we may sound upset – maybe we are; what we can say though is that we are working towards reconciliation. For these experiences are such a closeness of history that it and many Indigenous people are still living out the repercussions in their families and communities today. That is not to say we are unforgiving, we forgive. Though, in the true nature of what we are trying to accomplish here in this chapter, if you are offended by the directness of our thesis, that is fine, we understand. We only ask that you please remember this is the truth-telling that restorative justice needs, the awareness that slow violence demands, and we, meaning our perspective and individual stories, are only a very small part of the whole. But, if you are not going to participate in truth-telling then please step aside and do not hold it back. Perhaps this is not the best way or approach to bring resolution or even awareness to the issues, but it is one way.

Learning opportunities for justice

South and North Dakota are states where small-scale, community-based learning opportunities of justice born from injustice and the ongoing struggle of racial tensions could help to inform national forums for action. At the very least, the rest of the country (even the world) can learn from their continued struggles, and triumphs.[1] Deemed by some as two of the most racist states towards Indigenous peoples, the Dakotas up until 2016 proudly bolstered place names like *Harney Peak* on the landscape, after a military figure who committed acts of genocide against Indigenous people (Mark 2015). Indigenous people make up 29% of the total incarcerated population with a total state-wide population of 8.8% in South Dakota (U.S. Census 2010), and 20% of the total incarcerated population with a total state-wide population of 5% percent in North Dakota (Vera 2019). The Dakotas are places where U.S. soldiers received medals of honor for chasing down and murdering peaceful unarmed elders, children, women, and men (Abate 2010). They are places where large gatherings of peaceful Indigenous people and allies protesting to protect their rights to clean drinking water were brutalized and mistreated by authorities protecting corporations' capital interests. They are places where the courts have demonstrated a lack of care and compassion when determining measures of justice for 57 Lakota children, victims of a mass hate crime (Brave NoiseCat 2015). The following examples of current struggles for justice that Lakota youth face in the Dakotas serve to illustrate the core arguments of this chapter: that racist behavior perpetuates itself whether intentional or not, and that younger generations of Lakota youth are still dealing with the same violent issues of their great-grandparents' generation.

On January 25, 2015, a group of 57 Lakota children from the Pine Ridge reservation were attending a hockey game in Rapid City, South Dakota. The hockey trip was a reward for the children by their school for hard work and commitment to their education. It was reported that during the game a white fan, sitting in an executive suite above where the children sat, spilled beer on them while others in the suite verbally abused them with racist taunts. After the incident, school chaperones gathered the children and escorted them out of the building to their buses. The fan, who admitted consuming alcohol for hours prior to the incident, was later found not guilty of these allegations by a judge in a Rapid City, South Dakota court, and any pursuit of a hate crime was dismissed. The fan had indicated he was intoxicated and did not realize his beer was spilling on the children as he celebrated with a roping like motion with alcohol in hand after the team scored. Prior to this incident, it was reported that another white male in the suite was seen acting as if he was going to spill his alcohol on the children. The allegations of racist comments were dismissed after investigation by the Rapid City Police Department. Analysts of the event declared that the argument was ill-prepared and derived from a thin argument brought by the city that represented the children, which essentially spoke to a lack of interest to truly investigate the events (Heidelberg 2015; Giago 2018). Numerous rumors circulated about the students' behavior at the game, in what seemed to be a clear societal bid to discredit and shape the narrative in favor of the accused and thus depict how Lakota people/youth should be viewed by the citizenry. For example, to add insult to injury to the students, their families, and the community at large, the *Rapid City Journal*, a local newspaper that later apologized for publishing this hearsay, headlined "Did Native students stand for National Anthem?" six days after the incident, indicating an obvious attempt to obfuscate the story and blame the victims. Post and ongoing protests led by supporters of the children, and many others, such as their parents, directed attention towards this event and the deep underlying racial issues in Rapid City, South Dakota.

Situations like this in border towns surrounding reservation communities often spark a series of common responses in the aftermath of racist acts. Schools and communities initiate new programs, form boards, start commissions, hire experts, create positions like Native Student Councilor at local, predominantly white schools as a reaction to address and suppress racism. The transformative hope, at least at the core, is to educate school-age children in hopes of suppressing future racist behavior and to support the education as well as the safety of young Indigenous children in state schools.

While capacity building to address racism is an approach to be commended, the longevity and overall effectiveness of these programs, commissions, and boards is something to keep one's eye on. Beyond that, what about those 57 children? In full consideration of their experiences what pain do they carry? Do they feel safe and protected when in Rapid City? How can we begin to peel back the layers of confusion, pain, and anger they may feel as adolescents? We might expect that they may feel that justice, in terms of the authority that police officers and judges hold and the human rights they take oaths to protect, is not meant for Lakota children, but is meant for those who abuse them. Children who experience this behavior first-hand may feel physically and emotionally assaulted, and in the end justice for them is elusive, neither restorative nor reconcilable. Is it possible, then, for society to consider the slow violence associated with this event the children may carry? And for how long? The experiences of youth can stick with them their entire lives; while some may forgive, this particular violence can be slow, and that experience may be handed down to their children and family as well as community members. It is as Bryant-Davis et al. (2017) explain in the *Trauma lens of policing against racial and ethnic minorities*:

> According to the Diagnostic and Statistical Manual of Mental Disorders (DSM-5), the traumatized person may be the direct victim of the act of aggression, may witness or learn that it has happened to someone close to him or her, or may be repeatedly exposed to the details of the event (American Psychiatric Association, 2013). Psychological trauma may result in posttraumatic stress disorder (PTSD; intrusive thoughts, avoidance, and hypervigilance) but it is also associated with depression, distrust, affect dysregulation, panic, substance dependence, selfharming behaviors, shame, and difficulty focusing and functioning (Bryant-Davis & Ocampo, 2005) … we define racially motivated police brutality trauma as an act of violence or the threat of violence perpetrated by police officers against racial or ethnic minorities. Ethnic minorities who have experienced police brutality, directly or indirectly, may think about these instances when they do not want to think about them (nightmares, flashbacks, etc.) … and remain in a psychological state of high vigilance, on guard against the possibility of abuse … (Aymer, 2016).

Those who come to know trauma in its many forms know that it is a companion that can wield a heavy blow throughout one's life. Unfortunately, it is now solely the responsibility of their community, families, and allies to address the trauma children may carry from this event. To hopefully interrupt and heal the wounds of slow violence in the form of trauma, not only stemming from the incident, but the failure of justice itself to act in humane ways that reflect care and compassion, therefore demonstrating why the youth should believe in the system.

For the betterment of society as a whole, Lakota families and communities, like all U.S. citizens, are asked to help their children reconcile with justice, to believe in it. Even though their experiences with it as children may be painful, they should have faith in the justice system and the authority figures who represent justice. Essentially, the learning opportunity in this circumstance we would like to bring forward is a test of morality, not necessarily a legal challenge to convict and sentence those accused. More so, those who control the narrative (media writ-large) and actions (legal authorities) of caring and compassion in a political and legal system that failed to do so for these children handled this situation in a way that reflects the values of those institutions. To be clear, though, we are not in the position, nor do we want to be, to place judgment on another. However, we do want to point out the wrongs committed, and how not addressing these wrongs further disregards a sense of care and compassion, perpetuating inequity. In keeping with Indigenous forms of justice, we acknowledge that justice in this case should strive for an inclusive reconciliation for all.

The nature of this event, and events like it, do not occur in a vacuum. For many, they have become normalized, a part of the societal fabric in the Dakotas for white-Indigenous relations and the authorities who preside over these types of conflicts. While there are many non-Indigenous in both states who support Indigenous people, and vice versa, when prosecutable and egregious acts against Indigenous people occur, the system and those who preside over it are for many inherently flawed. Like many of the states in the U.S., systemic racism can be easily drawn out of readily accessible demographic data in the public preview. Following the unwarranted and frankly crushing deaths of numerous unarmed, non-threatening human beings of diverse minority and ethnic backgrounds leading up to and at the time this chapter was being written (2020), national statistics are becoming an increasingly sought-after story board to help shape a new unaltered and real community narrative that is written by, for, and with those who are experiencing systematic racism and racial injustice, which helps society work more towards awareness and change such as the study cited here:

> To get a clearer picture, Mike Males, senior researcher at the center on Juvenile and Criminal Justice, looked at data the Centers for Disease Control and Prevention (CDC) collected from medical examiners in 47 states between 1999 and 2011. When compared to their percentages of the U.S. population, Natives were more likely to be killed by police than any other group ... Males' analysis of CDC data from 1999 to 2014 shows that Native Americans are 3.1 times more likely to be killed by police then white Americans.
> Over the past 40 years, the U.S. Commission on Civil Rights (USCR) ... has held numerous hearings on discrimination in border towns surrounding reservations: ... in South Dakota, near the Sioux reservations. (Woodard 2016)

For instance, only a year after the "Lakota 57" incident, events that took place on the Standing Rock reservation, which borders both South and North Dakota just 200 miles northeast of Rapid City, illustrate flawed and racist police practices. Energy Transfer Partners, a company working in gas and oil, forcibly constructed an oil pipeline over Lakota lands and crossing the Missouri River. It is contended that along the way it broke countless tribal, federal, and state laws. After being denied permission to build and route the pipeline north of Bismarck, North Dakota due to the possibility of the pipeline breaking and affecting Bismarck residents' drinking water from the river, Energy Transfer Partners reportedly went to the tribe for permission. The tribe reportedly denied their request, citing similar concerns as those expressed by Bismarck residents, among other important cultural and spiritual reasons. Energy Transfer Partners ignored tribal documentation of long-established sacred tribal cultural sites and started bulldozing for pipeline construction, desecrating these sites along the way (Hand 2016).

What ensued was and continues to be a civil rights movement. The youth on Standing Rock took a leadership role to protect *Mni Wiconi*, the Lakota epistemological commitment to their relationship with water as one of the cornerstones of life. At its core, the Standing Rock movement was about the Lakota belief that water is a core purveyor of all life. Youth from Standing Rock organized a run to Washington D.C. in an act of solidarity to protect water and educate the public on the threats the pipeline posed to the drinking water supply of 18 million people spanning numerous states on the Missouri River. The youth, working with the tribe and allies, helped to lead a massive (approximately 10,000–15,000 people) non-violent, direct-action, water protection campaign with people from all over the world who traveled to Standing Rock. The record, in the form of video and legally documented testimony, shows the brutalization of those who protected the water using non-violent direct action by those who protected the interests of the company. This incident provided a gross display of what greed can and is willing to do to humans who oppose it.

Given the violence that was allowed, and that the position of the Obama administration at the time was "we're gonna let it play out for several more weeks" (Obama 2016), for some it was/is safe to assume those who opposed the construction were not eligible at the time to have their civil rights protected under constitutional law. Coupled with a short-term fix to temporarily deny the easement for the pipeline, which briefly stopped construction (Dennis and Mufson 2016), the reported violent actions taken by Morton County Police, and Energy Transfer Partners who hired security company TigerSwan, was considered acceptable given the outcome (Estes 2019). In reality, these events – the violent actions and the pipeline construction – could have been slowed or stopped completely. When the Trump administration took office one of the first orders of business was an executive action to grant the easement to Energy Transfer Partners (Almasy 2017). The violence inflicted on those who were unarmed and non-violent was reportedly horrific. Much of it was captured on numerous live and recorded video feeds showing altercations perpetuated by the Morton County police department who, in partnership with Energy Transfer Partners, i.e. TigerSwan, jailed and reportedly beat many of the demonstrators (National Lawyers Guild 2017).

Interestingly, the Morton County Sheriff at one point indicated that the United States Army Corps of Engineers requested their assistance to remove demonstrators from Corps lands, but this claim was later proven to be false (Nienaber 2016). At the time of this writing, North Dakota Governor Doug Burgum announced "Dakota Access Pipeline LLC has donated $15 million to the North Dakota Department of Emergency Services (DES) to help retire debt incurred by the state as a result of its response to the Dakota Access Pipeline protests," and North Dakota Senator, and "member of the Senate Appropriations Committee," John Hoeven "announced that North Dakota will receive up to $15 million in federal funding to help reimburse the state for costs incurred as a result of the Dakota Access Pipeline protests" (NDOG 2017; USSND Hoeven 2017). To some outsiders, the events that took place at Standing Rock demonstrate an outright violation of basic human rights, and to other outsiders those events were shaped by many media narratives as Indigenous people getting in the way of progress. It seems, though, that for many Lakota people who live in South and North Dakota this was business as usual, and a further strengthening of settler-colonial ideologies. In other words, both of these examples might demonstrate for Lakota youth that justice in the Dakotas does not include Lakota people or their allies. While the layers of slow violence are both inherently tangible as well as intangible, the work is to understand, at the very least in an abstract way, how events like this only add to the layers that need to be peeled back in order to fully understand how the legacy of violence and thus trauma continues to affect marginalized peoples. Such is the lived reality of those who have violent actions taken on them for non-violent direct action. Though it may not be the intent of those who have these experiences to hand down the feelings associated with slow violent action taken on them – such as anger, anxiety, fear, trauma, sadness, among many others – to younger generations, if society can commit to *seeing* their actions maybe then society can commit to *feeling* empathy. The internal question every human can ask themselves then is do I feel empathy for those who have been subjected to racial injustice, particularly violent acts on youth; then, why or why not. If the answer is *"no, or yes I feel empathy, but I am not implored to do anything about it, I am sure our law and policy institutions will work it out in a fair manner,"* this may be a conditioned behavior. A reflection of what Dr. Martin Luther King called the "white moderate" – the majority of this nation's white population that sees injustice but chooses to let others or those institutions who have been proven to be racially charged to lead. Building an inclusive positive future and normalizing empathy and therefore creating commonality may be where the real chance to make change lies.

Like the Lakota 57, the aftermath of Standing Rock will be something to keep our eye on, not only for the court rulings, of which thus far a few have been in favor of the tribe, and social programs developed in response to these events, but for the ongoing racial tensions in the Dakotas, carried out by the very public officials sworn to protect their citizens. While non-Indian populations continue to decline and Indigenous populations rise in the Dakotas, will justice still be controlled by the very few and out of reach for the many? On the national level, can we learn from the mistakes of the past, to provide for a better future? Let's be mindful, the Lakota 57 and events at Standing Rock involved Lakota youth. This is a part of the continued legacy of justice for entire generations of youth in the Dakotas. To revisit the core questions of this chapter: How does racism perpetuate itself in slow and violent ways? How do we as a society give precedence to racism by saying these events and those who preside over the legal questions that stem from these events are reasonable, and no societal change is taken, and no action is needed? There are too many layers to peel back in order to fully understand the political discourse of race, ethnicity, socioeconomic class, and sexuality at play here, but one thing is certain, these are the experiences of youth, and this then is about the generational experiences of the youth. The question of whether the traumatic events of one generation tend to manifest in future generations' lived experiences is no longer a debate: they do. While some may argue that those events happened a long time ago, they were isolated, those populations should "get over it," cast these events to the annals of history, and therefore in no way do generations of Lakota youth today have similar experiences as their great-grandparents. The task that lies ahead is working to understand how Indigenous youth are impacted. If the moral roots of justice are found in the fleeting practice of natural law – i.e., caring and compassion, to leave the world in a better place than it was left to us – our time could be spent addressing these issues so all youth can learn to be better human beings to all life.

Conclusion

Ideally, the transformative approach to justice, or the study of it, could be about action and change. It could be about bringing compassion to bear on difficult, deep-seated prejudices to uncover the taken-for-granted beliefs rooted in settler-colonialism. Inclusive approaches to justice seek to decolonize society, not only for the benefit of Indigenous communities and our youth, but also for settler-descendant communities as well. The survival of all is dependent upon the erasure of the harmful rhetoric of colonial structures of thought including in politics, science, and law. This is the true aim of decolonization all must acknowledge, but particularly those who are the descendants of settlers. Committing to decolonization is an immediate necessity in the spirit of Indigenous youth, but it is also crucial to dismantling the settler-colonial structures that slowly erode communities and the environment. Transformative justice requires difficult dialogues that expose the structures of domination and division inherent in the policies that underpin settler-states.

Sometimes, it is painful to be Indigenous in a nation that works to marginalize our lived experiences, much in the same way that other ethnic and racial groups are marginalized through slow, systematic processes of violence. To think, and in some instances to know, that there are people and organizations, including our legal institutions, that see Indigenous children as unworthy of reconciliation and justice is unconscionable. At times, the nation's tolerance of racist actions, couched or for some hidden within the Constitution and laws that guide our country's moral progress, hinders the ability to move forward as a nation, allowing hate and violence to slowly fester and our children to learn this kind of fear. Justice, whether reconciliatory, transitional, or restorative will only be realized when society, and the institutions society puts trust in, love all children more than the political discourse of capital, greed, race, ethnicity, socioeconomic class, and sexuality.

Note

1. For a more in-depth Lakota perspective of reconciliation, transformative and restorative justice in South Dakota see Edward C. Valandra (2005) "Decolonizing 'Truth': Restoring More than Justice," in Wanda McCaslin (ed.), *Justice as healing: Indigenous ways*, St. Paul, MN: Living Justice Press, pp. 29–53.

References

Abate, M. A. (2010), "'Bury My Heart in Recent History': Mark Twain's 'Hellfire Hotchkiss,' the Massacre at Wounded Knee, and the Dime Novel Western." *American Literary Realism*, **42** (2), 114–128.

Almasy, S. (2017), *Dakota Access Pipeline: Army issues final permit*, accessed February 8, 2017 at www.cnn.com/2017/02/07/politics/dakota-access-pipeline-easement-granted/index.html

Aymer, S. R. (2016), "I can't breathe: A case studying-helping Black men cope with race-related trauma stemming from police killing and brutality." *Journal of Human Behavior in the Social Environment*, **26** (3–4), 367–376.

Brave NoiseCat, J. (2015), "No one held accountable for Native kids harassed at South Dakota hockey game." *Huffington Post*, 9 February, 2015, updated January 2017.

Brewer II, J. P. and E. A. Kronk Warner (2014), "Protecting Indigenous knowledge in the age of climate change." *Georgetown International Environmental Law Review*, **27** (4), 585–628.

Bryant-Davis, T. and C. Ocampo (2005), "The trauma of racism: Implications for counseling, research, and education." *The Counseling Psychologist*, **33** (4), 574–578.

Bryant-Davis, T., T. Adams, A. Alejandre, and A. A. Gray (2017), "The trauma lens of police violence against racial and ethnic minorities." *Journal of Social Issues*, **73** (4), 852–871.

Deloria, Vine and Daniel Wildcat (2001), *Power and place: Indian education in America*. London: Fulcrum Press.

Deloria, Ella Cara (1998), *Speaking of Indians*. Lincoln: University of Nebraska Press.

Dennis, B. and S. Mufson (2016, December 5), "Army Corp ruling is a big win for foes of Dakota Access Pipeline." *The Washington Post*, accessed September 5, 2017 at www.washingtonpost.com/news/energy-environment/wp/2016/12/04/army-will-deny-easement-halting-work-on-dakota-access-pipeline/?utm_term=.58602aad3083

Estes, Nick (2019), *Our history is the future: Standing Rock versus the Dakota Access Pipeline, and the long tradition of indigenous resistance*. London and New York: Verso.

Giago, T. (2018), "Tim Giago: Apology owed for racist treatment of Native youth," accessed November 8, 2020 at https://www.indianz.com/News/2018/11/26/tim-giago-apology-owed-for-racist-treatm.asp

Gray, B. and P. Lauderdale (2007), "The great circle of justice: North American indigenous justice and contemporary restoration programs." *Contemporary Justice Review*, **10** (2), 215–225.

Hand, N. (2016), "Standing Rock Sioux Tribe Condemns Destruction and Desecration of Burial Grounds," accessed at https://indiancountrytoday.com/archive/standing

-rock-sioux-tribe-condemns-destruction-and-desecration-of-burial-grounds
-tbGDUq4PW0aOVUZEiVIweA

Hiedelberg, C. A. (2015), "Judge: O'Connell threw beer but didn't mean it, did not throw racial insults," accessed November 6, 2020 at https://dakotafreepress.com/2015/09/02/judge-oconnell-threw-beer-but-didnt-mean-it-did-not-throw-racial-insults/

Larsen, Soren C. and Jay T. Johnson (2017), *Being together in place: Indigenous coexistence in a more than human world*. Minneapolis: University of Minnesota Press.

Little Bear, Leroy (2000), "Jagged worldviews colliding," in Marie Battiste (ed.), *Reclaiming Indigenous voice and vision*. Vancouver: University of British Columbia Press, pp. 77–85.

Mark, Jason (2015), *Satellites in the High Country: Searching for the wild in the Age of Man*. Washington, DC: Island Press/Center for Resource Economics.

Marshall, T. F. (1999), "Restorative justice: An overview." A Report by the Home Office Research Development and Statistics Directorate. London: HMSO.

McAlinden, A. (2011), "'Transforming justice': Challenges for restorative justice in an era of punishment-based corrections." *Contemporary Justice Review*, **14** (4), 383–406.

McGregor, Deborah (2009), "Honouring our relations: An Anishnaabe perspective," in Julian Agyeman, Peter Cole, Randolph Haluza-DeLay, and Pat O'Riley (eds.), *Speaking for ourselves: Environmental justice in Canada*. Vancouver: University of British Columbia Press, pp. 27–41.

McGregor, D. (2012), "'Traditional knowledge: Considerations for protecting water in Ontario." *International Indigenous Policy Journal*, **3** (3), 1–21.

Mohawk, John (1992), "The Indian way is a thinking tradition," in José Barreiro (ed.), *Indian roots of American democracy*. Ithaca, NY: Akwe:kon Press, Cornell University, pp. 20–29.

Mohawk, John (2010), *Thinking in Indian: A John Mohawk reader*, edited by José Barreiro. Golden, CO: Fulcrum.

National Lawyers Guild (2017), accessed October 3, 2017 at www.nlg.org/water-protector-legal-collective-files-suit-for-excessive-force-against-peaceful-protesters/

Nienaber, G. (2016), "Morton County Sheriff blames USACE for police violence at Standing Rock as journalist is shot. *Huffington Post*, accessed 4 November 2016 at www.huffingtonpost.com/georgianne-nienaber/morton-county-sheriff-bla_b_12797812.html

North Dakota Office of the Governor (NDOG) (2017), accessed January 3, 2017 at www.governor.nd.gov/news/dakota-access-donates-15m-pay-down-loans-related-pipeline-protest-response

Obama, Barack (2016, November 1), "Now this Exclusive," accessed November 12, 2020 at https://www.nytimes.com/2016/11/03/us/president-obama-says-engineers-considering-alternate-route-for-dakota-pipeline.html

United States Senator for North Dakota John Hoeven (2017), website, accessed October 3, 2017 at www.hoeven.senate.gov/news/news-releases/hoeven-secures-funding-to-help-with-dapl-protest-costs

Vera Institute of Justice, Incarceration Trends in North Dakota (2019), accessed November 12, 2020 at www.vera.org/downloads/pdfdownloads/state-incarceration-trends-north-dakota.pdf

Weslager, Clinton Alfred (1989), *The Delaware Indians: A history*. New Brunswick, NJ: Rutgers University Press.

Whyte, Kyle P., Nicholas J. Reo, Deborah McGregor, M. A. (Peggy) Smith, James F. Jenkins, and Kathleen A. Rubio (2017), "Seven Indigenous principles for successful cooperation in Great Lakes conservation initiatives," in Eric Freedman and Mark Neuzil (eds.), *Biodiversity, conservation and environmental management in the Great Lakes Basin*. New York: Routledge, pp. 182–194.

Woodard, Stephanie (2016), "The police killings no one is talking about." *In These Times*, accessed September 23, 2017 at inthesetimes.com/features/native_american _police_killings_native_lives_matter.html

Yazzie, Robert and James Zion (1996), "Navajo restorative justice: The law of equality and justice," in Burt Gallaway and Joe Hudson (eds.), *Restorative justice: International perspectives*. Monsey, NY: Criminal Justice Press, pp. 144–151.

10 The infliction of slow violence on first wives in Kyrgyzstan

Michele E. Commercio

Introduction

"I have to be silent."

Zarina, a woman who participated in a focus group with first wives in Kyrgyzstan, described her marital history as follows: "I think first about myself and then about my children. I depend entirely on my husband financially. Without him … Well, I married early, I have no education – where can I go? I'm already forty years old, so I have to be silent. What's there to do? … I tolerate it because there's no way out … It's distressing because I'm a beautiful, good woman … what else did he need?" Zarina endures the fact that her husband has a second wife because she is financially dependent on him, but she feels trapped in a painful familial situation she struggles to comprehend. That sense of being stuck in a troubling circumstance is a manifestation of the slow violence Zarina's husband inflicted on her when he took a second wife.

This chapter examines slow violence in the family by considering potential impacts of polygynous marriages on first wives and children growing up within these unions in Kyrgyzstan, where polygyny is classified as a crime against the family and minors. Article 179 of the Kyrgyz Criminal Code punishes polygamists with public service, wage garnishment, or a fine.[1] I argue that the act of taking a second wife can represent a form of slow violence imposed on first wives, who have *limited* response options because of socially constructed constraints on women, and on children of first and second wives, who experience adverse consequences of polygyny but as minors have *no* response options. This chapter also considers the connection between slow violence and place by taking into account economic and political factors that have led to a flourishing of polygamous marriages in post-Soviet Kyrgyzstan.

Rendering an invisible phenomenon visible

Johan Galtung distinguished manifest violence, which is overt, explicit, and observed, from latent violence, which "is not there, yet might easily come about," in order to develop his theory of structural violence (Galtung, 1969, p. 172). Building on Galtung's work, Rob Nixon introduced the term slow violence eight years ago, when he argued for the need to view critical environmental challenges like climate change in terms of a gradual, invisible manifestation of violence. In contrast to typical constructs of violence that focus on explosive events, slow violence "occurs gradually and out of sight" and is "neither spectacular nor instantaneous, but rather incremental and accretive, its calamitous repercussions playing out across a range of temporal scales" (Nixon, 2011, p. 2). Emphasizing the delayed effects of slow violence, Nixon builds on Galtung's work by expanding the notion to include more nuanced understandings of violence inflicted slowly over time. A husband's decision to take a second wife in Kyrgyzstan is an act of slow violence if it causes women and children to suffer over time. Although I focus on women who oppose their husbands' decision to take a second wife, it is important to note that there are women who either initiate or support this decision for various reasons.

I propose a qualitative, multi-methods approach for investigating this act of slow violence, or for rendering this form of slow violence visible. To address the complexities associated with researching an illegal activity with invisible repercussions, I obtained as many varied data points as possible and adopted an ethnographic approach that, as Edward Schatz recommends, focuses on the understandings subjects – in this case first wives – have of their own lived realities (Schatz, 2009). Data sources thus include: (1) semi-structured interviews with people from different sectors of Kyrgyz society including former and current government officials, civil society, academia, and the intelligentsia; (2) focus groups with first wives and focus groups with men who have multiple wives residing in the capital; (3) Russian language news articles about polygyny in Central Asia; (4) Kyrgyz legislation pertaining to the practice; and (5) relevant secondary sources. I conducted interviews to hear stories from individuals who encounter first wives in their professional lives and focus groups to hear stories from first wives and men with multiple wives. Target interviewees were chosen on the basis of purposive sampling to tap into the expertise of locals who work with women's issues as they are defined in Kyrgyzstan. I have changed the names of all interviewees in order to ensure anonymity. Data gathered during trips to the field in 2013, 2015, and 2018 inform the argument made in this chapter.

In 2014, I hired a Kyrgyz research firm called *El-Pikir Center for Public Opinion Study* to conduct two focus groups with first wives. Participants had to be Bishkek residents and at least 18 years old. I sought through these focus groups to glean insights into purposes and understandings of polygyny, to have an alternative data source with which to compare individual interview data, and to generate from this comparison more trustworthy findings. Twenty women participated in the first wives' groups (ten per group); the age range was 31 to 65, and the average age was 45. In 2018, *El-Pikir* conducted three focus groups with polygamist men. Participants, who had to be Bishkek residents and at least 18 years old, also had to consider themselves married to, rather than simply involved with, two women. Twenty-seven men participated: eight in one group, nine in another group, and ten in yet another group. The age range was 25 to 64, and the average age was 44. One participant had three wives; the rest had two.

To analyze focus group data in terms of individual and group perspectives I took or constructed from each discussion marital stories told by participants about themselves, their relatives, or their acquaintances, and analyzed exchanges between participants that revealed points of agreement and disagreement. Rather than assign either data source more weight, I integrate focus group and individual interview data by analyzing consistencies and inconsistencies from both sources. This chapter relies on the "arresting stories" of first wives and individuals qualified to speak authoritatively about polygyny as an indicator of slow violence inflicted on women by men who embody a patriarchal culture that permeates society and a state that is unable and/or unwilling to address this issue.

The context of today's arresting stories

Nixon suggests that the representation of slow violence is possible through the telling of "arresting stories," even if they are incomplete (Nixon, 2011, p. 3). Alima told her focus group peers the following incomplete but arresting story: "Earlier, before they closed the [Communist] Party, it was forbidden for him [my husband] to marry [a second wife]. Then I gave my consent so that he would help me and the children. I had no other way out … I wouldn't want this for my enemy." Although the Kyrgyz practiced polygyny prior to the Soviet era, Alima's story suggests that the Communist Party limited its frequency. While the Party was unable to eliminate 'vestiges of the past' like bride theft and polygyny, it did curb the prevalence of these practices (Chirkov, 1978,

p. 197). The Party's responses to immoral behavior included criminal prosecution, expulsion, and public shaming.

Court cases heard during the Soviet era in the Kyrgyz Republic resulted in convictions of men with multiple wives like T.A., who was sentenced to one year of corrective labor because he took a second wife in 1949 – 19 years after he married his first wife (Kozhonaliev, 2000, pp. 147–148). Party expulsion and public shaming were, however, more common than criminal prosecution. As a local expert on gender issues explained:

> During the Soviet era there was criminal punishment, it was a criminal act ... Any woman who had a suspicion that there was such a family [polygynous] could simply complain to the professional union at work or to the Party. It was strictly controlled. I assume that people walked a path that lay under this control, and that polygyny was limited to isolated cases and wasn't flaunted.[2]

A Kyrgyz historian with a long-term party career confirmed the Party's success in containing polygyny:

> There were Party Commissions in the factory that made decisions, they discussed cases openly. I worked in different party organs and when polygyny was detected, men were expelled from the Party ... A wife would submit a report, the Commission would examine it, and then the Party would expel him and he wouldn't be able to work in that sphere again. If the wife didn't submit a report, nothing changed; if she did, that was it.[3]

Focus group data illustrate the Party's role in discouraging the practice in the Kyrgyz Republic. The consensus among first wives was that there were few cases of polygyny during the Soviet era because the Party considered it immoral and reacted severely to men with multiple wives. The following exchange among first wives suggests that the Party gave women power to prevent their husbands from inflicting this manifestation of slow violence on them, or to punish husbands who had already done so:

Elmira:	I remember that a wife could submit a report.
Moderator:	Yes, yes.
Elmira:	And then attend a meeting to discuss ...
Moderator:	A reprimand, right?
Elmira:	Right down to putting aside his membership card if he was Party.
Asel:	That's how it was.
Elmira:	They dismissed him from his position and covered him with shame from all sides.
Asel:	Yes, yes.

According to a representative of a local Islamic organization, a Party member's wife had the power to at least discourage, if not prevent, her husband from taking a second wife:

> My husband was a member of the Communist Party. I told him all the time "You're Party, if you take a wife I'll put you in jail. I'm a *Komsomolka*. And you are Party." If a person was Party and had a second wife, they kicked him out. In general polygyny didn't exist because if someone heard about it, they'd expel you from of the Party.[4]

The rise of polygynous marriages in contemporary Kyrgyzstan is connected to the Soviet Union's collapse, which created immense political, economic, and social upheaval. This particular manifestation of slow violence is flourishing within the context of an ongoing turbulent economic transition that is not coupled with an official ideology based on a clearly articulated moral compass. Kyrgyzstan's political transition rendered the Communist Party – and its ability to restrict what it deemed immoral behavior – a remnant of the past. This had two consequences for women: first, it deprived them of the power they had during the Soviet era to prevent their husbands from taking second wives; second, it created an ideological vacuum that is characterized by some degree of societal Islamicization. The end result, intended or not, was the emergence of a society in which men have a religious justification for taking a second wife – Quranic verse 4:3 – and an institutional mechanism that enables this act in a legal context that criminalizes polygynous marriages – *nikah* – while women lack the power to oppose a husband's decision to commit this act of slow violence.[5]

Kyrgyzstan's economic transition created a pool of women who might agree to wed a married man in order to obtain marital status, the opportunity to reproduce, and/or economic security. During the 1990s, many young women put marriage on hold. In response to the dire economic situation, privileged young women went abroad to study while ordinary young women sold goods at the bazaar in order to feed their families. Social norms regarding the importance of marriage and motherhood put these single women in a precarious position as they entered their late twenties and early thirties. These widespread social norms, a shortage of eligible bachelors due partially to labor migration, and economic inequality between men and women led and continues to lead single women to contemplate second wife status, which is no longer condemned by an overarching political institution or political ideology.

Slow violence and children of polygynous marriages

In order to curb early marriages and polygyny, legislators have attempted to introduce legislation that would require a couple to possess a state marriage license before initiating a *nikah* ceremony. When I asked a government official familiar with these efforts about polygyny, he/she argued that the practice inflicts slow violence on children: "First and foremost, children suffer. And it violates the rights of women."[6] The adverse effects of polygyny on children came up repeatedly during the course of my research. Following Sheridan Bartlett, who draws a porous boundary around suffering to include not only abuse and physical force but also neglect and emotional harm, I argue for a loose interpretation of slow violence that also includes financial hardship (Bartlett, 2018, p. 5).

Financial hardship is one possible manifestation of slow violence inflicted on children born to first or second wives. Although Quranic verse 4:3 requires husbands to support all of their wives equally, some men in Kyrgyzstan find it difficult to meet this expectation and, as a result, cause their children to suffer. Men who participated in focus groups underscored this problem when asked about disadvantages of polygyny. One participant regretted taking a second wife precisely because of this Quranic stipulation: "If I could return to that time, I would have only one wife ... there are many problems – it's hard to feed everyone, financially it's tough." Another interpreted drawbacks of polygyny as follows: "There won't always be sugar in life, sometimes it happens that you don't have money. You need to buy coal, clothing for the children, funerals, weddings, sometimes there's not enough money for this ... ." When the moderator of one group asked participants if they would advise their sons to take multiple wives, one polygamist emphasized financial challenges as the reason he would not do so: "No, I wouldn't advise it. It's necessary to try to live in one marriage so that everything there is good. Because two families – that's a big expense. You need to buy the children many things – school, school uniform."

When discussing polygyny, interviewees echoed Bartlett's emphasis on ambient violence and "the routine misery for children that results from the deprivation, frustration and humiliation of their families and neighbours" (Bartlett, 2018, p. 5). Routine misery can manifest itself in emotional turmoil, which can be more devastating than financial hardship in the long run. A man who takes a second wife may enact emotional violence on children born to his first *and* second wife. The child of a first wife, Bolot is a 30-year-old man from Kyrgyzstan who describes his childhood in terms of shame: "The most offensive thing was that our whole town knew about this, my school, my mama, even my older brothers. But I couldn't speak with him [my father]." Neither Bolot nor his brothers supported their father's decision to move in with his other family when their mother passed away (Iarmoshchuk and Zhetigenova, 2019).

Since first and second wives in Kyrgyzstan generally do not reside in one household, men with multiple wives live with one wife and visit the other periodically. Children of polygynous marriages thus see their father on a part-time basis. This, as a polygamist who participated in a focus group explained, can be difficult for them to understand: "It's hard for a child when he asks 'where's my papa?' Well, he's there with the other mama … ." In 2013, a women's organization focused on politics prepared a roundtable discussion featuring men who had been raised in polygynous families in order to assess the impact of polygyny on children. Participants agreed that the practice adversely affects children. According to a representative of this organization:

> These men discussed the moral situation in their family, how they were raised, and they opposed polygyny because they saw that when fathers come and go the situation is not good. They told me about this, they said "We are absolutely against it, we'll never marry a second time, we'll never run between families because this affected us negatively as children."[7]

A government official I interviewed underscored the negative impact of polygyny on children as well when he/she described polygyny as problematic for women *and* children:

> Polygyny is a problem because it negatively influences the consciousness of women, [sending the message] that it's normal, that such relationships are normal. For children it's also very bad because it doesn't happen without conflict. There's always conflict … The first wife in fact is against this, she is very angry, and her relationship with her husband deteriorates. Then when there are children relations become very complex … .[8]

Jealousy between children of first and second wives is one manifestation of this complexity. As another government official explained:

> There is always rivalry between children of one father and two mothers. And when that rivalry emerges, it affects their upbringing. Children experience hatred not only within the family, but also when they leave the family. Their upbringing has a very negative side, and I don't think this should be an example for us.[9]

That upbringing can be marked by emotional instability. When I asked a representative of a local human rights organization how polygyny affects children, he/she highlighted a child's inability to comprehend his familial situation in terms of emotional suffering: "Children don't understand why their papa comes once every two weeks, they don't tell the children that he he's gone to the other wife. The child sees this, and everyone suffers."[10] In the following quotation, a first wife who participated in a focus group describes the suffering of her children as a "very painful spot":

> I had a bad situation; my husband lives with me and his second family, we have two children and he has a third from the second wife. He once made good money, I didn't work, I stayed at home. I had plenty, but there was resentment because I raised two children who lacked a father. When they still played outside they would run to the house and say "Mama, why has papa come to us only once this week?" What kind of answer can you give a child? A woman becomes withdrawn and irritable from this. And the children suffer. This is truly a very painful spot.

A crisis center representative who is adamantly opposed to polygyny because it adversely affects the psychological development of children articulated a similar sentiment:

> Children aren't satisfied with their father because their mama's always crying; the child sees this and his resentment grows. He resents his mama, he's always cursing "why did papa abandon us, why is this my mama's fate?" So his resentment grows, and his future in society will be a tyrant. This is a problem – the child's psychological development … I am categorically against the legalization of polygyny because it violates the rights of children. Not because I am a woman, but because children are hurt, they constantly ask "when will papa leave, why does papa leave?"[11]

When asked about disadvantages of polygyny as currently practiced in Kyrgyzstan, participants in a focus group of first wives agreed that the practice adversely affects children. One participant summarized this view as follows: "Regardless, there's a shortage of paternal caresses for the children."

Surnames are critical in Kyrgyz society because they either represent a financial and social resource or handicap. Children born to second wives experience a particular form of slow violence if they do not have their biological father's surname. A father may or may not establish paternity, thus giving or withholding his surname at the time of birth. In the eyes of the state, paternity renders the man in question legally responsible for a child he has had out of wedlock. Not surprisingly, men marrying second wives often decline to establish paternity. In this case, the biological mother – the second wife – gives the child her surname. This has economic and social implications. According to a representative of a local human rights organization, children born to second wives suffer financially if they have their mother's surname because they do not have rights to their father's assets: "If the father registers the children of this wife in his surname he must divide his assets. The law protects the child with his father's surname; this child automatically has the right to an inheritance. Naturally, the father doesn't do this."[12]

In 2016, this issue prompted one member of parliament to promote the legalization of polygyny in order to ensure that children of second wives have their father's surname (Momunov, 2016). The surname issue remains acute, in part, because the legalization attempt failed. According to a representative of Kyrgyzstan's legal community, if a biological father legally "recognizes" children born to his second wife, they have claims to his inheritance; if he fails to do so, they do not have a leg to stand on in court.[13] A representative of a local women's organization focused on gender equality summarizes the problem below:

> There is a problem in the fact that when a polygamist has children in a second marriage, they are registered in the surname of the wife because the marriage isn't registered. Accordingly, he doesn't establish paternity. According to the documents, the father is a dash. His surname is recorded only if he wants it to be. Problems concerning the rights of children born to second wives arise from this – inheritance, financial security, child rearing and so on. And of course this is considered bad, immoral, and shameful.[14]

Possessing your mother's surname is a financial and social handicap; having a dash in documents where your father's name should appear is shameful. However, being the child of a *tokol*, or second wife, can inflict emotional harm even if the child in question has his father's surname. The Kyrgyz word *tokol* means helpless, or unprotected. According to a crisis center representative who prides herself on her ability to speak Kyrgyz fluently, the contextual meaning of the word *tokol* that underscores powerlessness is 'sheep without horns.'[15] A representative of a local human rights organization agreed with this sentiment: "*Tokol* is a degrading word in Kyrgyz society. It means woman without status. When we humiliate we say 'you are a *tokol*,' meaning a woman without status – a mistress or a second wife."[16] The Kazakhs, who say *tokal* rather than *tokol*, have a fairly defensive saying that translates as follows: "I was born from a *tokal*, so what?" (Iarmoshchuk and Zhetigenova, 2019).

A representative of Kyrgyzstan's think tank community highlights problems *tokol* children have precisely because the state does not recognize their mother's marriage: "There's a problem in the fact that our law doesn't recognize polygyny, and therefore we have problems with children and second wives regarding the right to inheritance."[17] Similarly, a representative of a local Islamic organization described children of second wives as "no one" according to the law because without established paternity they lack rights to their biological father's assets.[18] Although first wives are legally entitled to their husband's assets if their marriages are registered with the state, they have problems that can be viewed as manifestations of slow violence.

Slow violence and first wives in polygynous marriages

A civil society representative quoted in local newspaper *Vchernii Bishkek* argues that while first wives, second wives, husbands, and children suffer from polygyny, "First wives experience the most painful moral, psychological burden" ("Odnazhdy nash muzh…," 2017). This burden stems, in part, from a husband's failure to consult his first wife before taking a second wife. When discussing Quranic stipulations regarding polygyny, a representative of Kyrgyzstan's think tank community argued that 98 percent of the country's Muslims are not acquainted with what he describes as a whole system that includes a provision requiring men to consult their first wives.[19] One focus group participant, a practicing doctor, returned from a business trip and found her husband's new second wife and the woman's daughter in her house. In order to cope with this unexpected development, she began to live a life according to Islamic principles.

Though they usually reconcile themselves to their new marital arrangement because married status is superior to single/divorced status, it may be difficult for first wives to do so. A local expert on Kyrgyz elections, and specifically the role of gender in these contestations, describes this process of adjustment in terms of emotional discomfort:

> It seems to me that when a husband takes a second or third wife, the first wife usually experiences a very high degree of discomfort. A very high degree of discomfort because she got married, doesn't have a profession, already has children, and he takes another and abandons her. She's forced to agree to such conditions.[20]

Elmira, who participated in a focus group of first wives, was able to accept her husband's decision because her relationship with her husband was exhausted:

> Why is it easier for me to experience this? Because I have two daughters and a son. I was already at the age, ten years ago, when you have the sense that your relationship is exhausted. I was already tired. Therefore I accepted – I was not ecstatic – but I accepted [the second wife]. Well, it's love, he fell in love. What are you going to do?

Galina's story is similar to Elmira's story in the sense that she reconciled herself to the fact that her husband took a second wife after a year of not being able to stomach the situation. It differs from Elmira's story, however, in the sense that Galina's husband tried to conceal his second wife:

> I didn't even know he was married, this information came from his friend. He works for a construction company, he's constantly on business trips, he's rarely home, he goes to different places including his mother's. He disappeared for half a year, and as always I hoped he might be on a business trip or at his mother's. He didn't call, but he did communicate with the children via Skype when he was missing. And then his friend came over and said "You know that he got married over there." I was upset, worried that the children would be upset, cry, suffer. When he came home there was profanity, a row, we gave each other the eye ... it was miserable. For one year I couldn't stomach it, and then I learned to reconcile myself to the fact that he has a second wife.

Some first wives are less accepting and articulate through their marital stories emotions of ongoing distress. For example, Chinara experiences slow violence emotionally and economically because her husband took a second wife:

> You feel very bad when your husband takes a second wife. A simple example – I'd just given birth and left the hospital, but he didn't care about this and went to the other. Sometimes he lingers with us, and then that wife pesters over the phone and says "Why haven't you let him go?" Then you don't know what to do, you feel bad and ask yourself "Who defends my rights, to whom do I appeal?" ... If we talk about a triangle, then the rights of the first wife are infringed upon the most because the first wife tolerates humiliation from her husband *and* the young wife. When you tell him that he isn't going to her, he argues and even raises his hand. Then he squeezes the financial situation as if he wants to say "I provide for you, sit and keep quiet." (Elkeeva and Sarygulova, 2015)

Asel, who participated in a focus group with first wives, also described her current marital situation in negative terms although emotional violence is more pronounced for her than economic violence:

> He helps financially, but I have no relationship with him and what's there to do? I gave him my youth, my beauty, and then he took it, he crucified me, he dragged me through the mud ... I'm telling you that it's horrible. It turns out that I'm a robot who raises children and earns money honestly. At night I sleep with a pillow and tears. Sometimes depression sets in and I want a shoulder; I want to have an interesting person to talk to and love. He found love for himself there, a wife with whom he fights, makes up, kisses – but me? Who thinks about me?

While public discourse on polygyny in Kyrgyzstan tends to focus on second wives and their lack of rights, some of it highlights slow violence experienced by first wives like Asel. One civil society representative has argued for the need to think about first wives:

> Today they propose to defend the rights of second wives. But what about the rights of a first wife and her children? For example, a woman can make riches and significance out of a mediocre man, devote herself to his whole life. But then he abandons her as she grows old. (Korotkova, 2017)

A member of the intelligentsia familiar with the Kyrgyz film *My Father's Wife*, which depicts the suffering of a first wife, told me that he/she had witnessed the suffering of several first wives with situations similar to Asel's situation: "I saw this suffering, this incredible suffering that a woman who's been married for fifty years experiences, a woman who planned to grow old and raise grandchildren with her husband. It is an unbelievable pain … I wanted to show what this woman experiences … ."[21] Begimay, who participated in a focus group with first wives, echoed this sentiment when she spoke about the future she thought she and her husband would have: "I thought we'd have the life of a story, we'd live together, and die in one day, as they say… ."

One interviewee told me about a wealthy acquaintance who took a second and third wife and, in so doing, devasted his first wife: "The first wife simply grew old because it was so difficult. And the children suffered. Emotionally, it destroys the first wife … If she cannot make peace with it, she will die."[22] In addition to emotional slow violence, first wives may also experience economic slow violence if they are financially dependent on their husbands. According to a representative of a local women's organization focused on politics, economic dependence limits the options available to women who are confronted with a second wife: "The first wife cannot open her mouth because she is dependent on him financially … He says to his her 'either shut your mouth, be silent, and endure, or you and the children will be left with nothing.' What can this woman do? Yes, she's silent. She cannot say anything, she must endure."[23] A Kyrgyz Ombudsmen summarized this phenomenon when she stated that many first wives remain married because their husbands feed and clothe them (Elkeeva, 2017).

Women in Kyrgyzstan may be financially dependent on their husbands because they are uneducated and/or have not pursued a career, and because there is a growing tendency for men to register their assets with blood relatives – rather than their wives – so that in the event of divorce they are resource-poor on paper. A crisis center representative summarizes the interaction effect of these two factors on a first wife as follows:

> The first wife sat home, her children are grown, she's done everything for her husband, she hasn't had a career. When she's fifty she may end up on the street because all movable and immovable property is registered in his relatives' names. They say to her "Good riddance!'" … She, of course, suffers from emotional and economic violence.[24]

Although first and second wives visit crisis centers, the former tend to report problems associated with polygyny because of the emotional and/or economic violence they experience on a regular basis. One crisis center representative told me about a client who was thrown out of her house when her husband brought a second wife home:

> First wives come with psychological problems and because they are without a home, without clothing, they have nothing, not even food … A woman came to us, born in 1963, her children are grown, she's lived with her husband for twenty-one years. He is an invalid, sits in a wheelchair, blind. Nevertheless he brought a young wife into the house, and abandoned his first wife like an unnecessary object. She was left on the street, so she came to us for help. There are many such examples.[25]

Sabira, who married at a young age, is another example. When she and her husband returned home from a labor migration stint in Russia, her husband took a second wife. In reaction to this act of slow violence, Sabira tried to commit suicide and then turned to Islam for comfort. However, as the author of the article featuring Sabira points out, "She raises her three children and tries not to show that she suffers" ("Odnazhdy nash muzh…," 2017).

The perpetuation of slow violence in contemporary Kyrgyzstan

Sabira's decision to remain married to a man who took a second wife against her will is not uncommon. Socially constructed constraints on women, as well as the state's casual attitude toward polygyny, led some women to stay in polygynous marriages. Regardless of the choice they make, first wives have *some* agency as they strike what Deniz Kandiyoti calls patriarchal bargains, or strategies adopted within certain identifiable socially constructed constraints (Kandiyoti, 1988, p. 275). In Kyrgyzstan, discernable socially constructed constraints placed on women include pressure to marry, reproduce, and maintain the household while raising children. Held by men and women, these expectations have deep roots in Kyrgyz nomadic culture. As historian and scholar P. Kushner wrote in 1929, "The goal of marriage for every Kyrgyz is reproduction. Childlessness can serve as sufficient basis for divorce of a woman. It is *necessary* to have a progeny ... The appearance of a child ... is considered an unexpected and very valuable gift" (Kushner, 1929, p. 72). Today women are still expected to maintain the hearth, a job that includes household duties, childrearing, and facilitating their daughters' marriages by maintaining close family ties.

These constraints – marriage, reproduction, and maintenance of the hearth – represent the parameters of the patriarchal bargains that women negotiate in Kyrgyzstan. In addition to financial dependence, first wives who chose to remain married do so because of these constraints. Once their husbands have taken an additional wife, first wives have two *realistic* options that lead to different outcomes. A first wife can initiate divorce proceedings, but this leads to an insecure future. A *divorcée* loses rights to her husband's inheritance, assets that are registered in her husband's name, the esteemed status she had as a first wife in the extended family, respect in her community, and the means to survive if she is economically dependent on her husband. Divorce also puts a woman's unmarried daughters in a precarious position because girls from broken families are far less desirable than girls from intact families. Alternatively, a first wife can tolerate her husband's second wife, and in so doing possibly increase her chances of economic security. Women who remain married to men who have taken additional wives maintain their marital status and may retain financial security – resources that benefit themselves and their children.

Both options are suboptimal for first wives like the ones featured in this chapter. Women who remain in polygynous marriages and children of polygynous marriages may experience invisible manifestations of harm and destruction at the familial, societal, and state levels. The societal expectations held by men and women discussed above are firmly embedded in these arenas. While the Kyrgyz state designates polygyny a crime, it avoids prosecution: no one committing this crime has been punished. Journalist Nurzada Tynaeva asserts that Article 179 of Kyrgyzstan's Criminal Code has never been invoked: "Representatives of non-governmental organizations and lawyers cannot recall a single case of a first or second wife suing her husband for polygyny" (Tynaeva, 2016). This is hardly surprising; women worry about emotional, economic, and/or physical retribution, as well as abandonment. Moreover, it is difficult to prove that a man has a second wife because the Criminal Code defines polygyny as cohabitation and the keeping of a common household with two or more wives. Cohabitation is easy to conceal because polygynous marriages are consecrated solely via *nikah*, which means there is no state record of the marriage. The keeping of a common household is also easy to conceal because men tend to register assets in the names of blood relatives, which means there is no state record linking the man in question to property ownership.

On the rare occasion that the Kyrgyz state has been confronted with a public case of polygyny it has opted to avoid prosecution. The Zhalilov case illustrates my point. In 2017, an influential former Mufti announced in a YouTube video that he had taken a second wife ("Aida Kasymalieva ...," 2017). Though the video kicked off an intense public debate, the state did not prosecute Zhalilov. As a representative of a local association coordinating the work of women's organizations in Kyrgyzstan explained: "There is a criminal stipulation, but it doesn't work ... It's already to the point where men aren't ashamed to admit that they are violating the criminal code of the country and not being punished for it."[26] This form of slow violence inflicted on first wives and children cannot be ameliorated without a change in societal expectations regarding the role of women in Kyrgyz society *and* a change in the state's attitude.

Acknowledgments

I would like to thank IREX (International Research and Exchanges Board), the Louis Rakin Foundation, and the University of Vermont for generous funding that supported this research.

Notes

1. *Ugolovnyi Kodeks Kyrgyzskoi Respubliki*, http://cbd.minjust.gov.kg/act/view/ru -ru/111527/20?cl=ru-ru&mode=tekst.
2. Personal interview, 14 May 2018: Bishkek, Kyrgyzstan.
3. Personal interview, 10 May 2018: Bishkek, Kyrgyzstan.
4. Personal interview, 22 July 2015: Osh, Kyrgyzstan.
5. Quranic verse 4:3 allows a man to have a maximum of four wives *if* he can provide for and treat them equally, and *nikah* is the religious ceremony consecrating a marriage.
6. Personal interview, 23 May 2018: Bishkek, Kyrgyzstan.
7. Personal interview, 11 July 2015: Bishkek, Kyrgyzstan.
8. Personal interview, 21 May 2018: Bishkek, Kyrgyzstan.
9. Personal interview, 16 May 2018: Bishkek, Kyrgyzstan.
10. Personal interview, 25 May 2018: Bishkek, Kyrgyzstan.
11. Personal interview, 24 July 2015: Osh, Kyrgyzstan.

12. Personal interview, 25 May 2018: Bishkek, Kyrgyzstan.
13. Personal interview, 15 May 2018: Bishkek, Kyrgyzstan.
14. Personal interview, 23 July 2015: Osh, Kyrgyzstan.
15. Personal interview, 13 July 2015. Bishkek, Kyrgyzstan.
16. Personal interview, 9 July 2015: Bishkek, Kyrgyzstan.
17. Personal interview, 22 May 2018: Bishkek, Kyrgyzstan.
18. Personal interview, 9 July 2015: Bishkek, Kyrgyzstan.
19. Personal interview, 22 May 2018: Bishkek, Kyrgyzstan.
20. Personal interview, 25 May 2018: Bishkek, Kyrgyzstan.
21. Personal interview, 8 May 2018: Bishkek, Kyrgyzstan. The film *My Father's Wife* can be found at: http://culture.akipress.org/news:1392335
22. Personal interview, 8 May 2018: Bishkek, Kyrgyzstan.
23. Personal interview, 8 May 2018: Bishkek, Kyrgyzstan.
24. Personal interview, 24 May 2018: Bishkek, Kyrgyzstan.
25. Personal interview, 23 July 2015: Osh, Kyrgyzstan.
26. Personal interview, 23 May 2018: Bishkek, Kyrgyzstan.

References

"Aida Kasymalieva: Mnogozhenstvo obestsenivaet Kyrgyzstan kak svetskoe gosu-darstvo" (2017, 2 December), *Vechernii Bishkek*, accessed 9 September 2019 at http://vb.kg/society/aida-kasymalieva-mnogojenstvo-obescenivaet-kyrgyzstan-kak-svetskoe-gosydarstvo.html

"Аида Касымалиева: Многоженство обесценивает Кыргызстан как светское государство," *Вечерний Бишкек*: http://vb.kg/society/aida-kasymalieva-mnogojenstvo-obescenivaet-kyrgyzstan-kak-svetskoe-gosydarstvo.html

"Aida Kasymalieva: Polygyny lessens the value of Kyrgyzstan as a secular state."

Bartlett, Sheridan (2018), *Children and the Geography of Violence: Why Space and Place Matter*, London: Routledge.

Chirkov, P. M. (1978), *Reshchenie Zhenskogo Voprosa v SSSR 1917–1937*, Moscow: Mysl'.

П.М. Чирков, *Решение Женского Вопроса в СССР 1917-1937* (Москва: Мысль, 1978).

Resolution of the Women's Question in the SSSR 1917–1937

Elkeeva, Kanymgul (2017, 31 July), "Mnogozhenstvo. Zhivut, potomu chto on kormit i odevaet," accessed 9 September 2019 at https://rus.azattyk.org/a/28650201.html

Элкеева, Канымгул, "Многоженство. Живут, потому что он кормит и одевает," 31.07.2017, Радио АзаттыкЖ: https://rus.azattyk.org/a/28650201.html

"Polygyny. They live because he feeds and clothes them."

Elkeeva, Kanymgul and Burulkan Sarygulova (2015, 30 September), "Mnogoznenstvo – udar toporom po korniu naroda?" accessed 9 September 2019 at http://www.gezitter.org/society/43932_mnogojenstvo_-_udar_toporom_po_kornyu_naroda/

Канымгуль Элкеева и Бурулкан Сарыгулова, "Многоженство – удар топором по корню народа?" 30 сентября 2015, *Gezitter.org*: http://www.gezitter.org/society/43932_mnogojenstvo_-_udar_toporom_po_kornyu_naroda/

"Polygyny – an attack with an ax at the roots of the people."

Galtung, Johan (1969), "Violence, Peace, and Peace Research," *Journal of Peace Research* **6** (3), 167–191.

Iarmoshchuk, Tat'iana and Alina Zhetigenova (2019, 4 April), "Zachem muzhchiny v Tsentral'noi Azii zavodiat vtoruiu sem'iu," accessed 9 September 2019 at https://rus.azattyq.org/a/central-asia-polygamy/29863234.html

Татьяна Ярмощук и Алина Жетигенова, "Зачем мужчины в Центральной Азии заводят вторую семью," *Радио Азаттык* 4/4/2019: https://rus.azattyq.org/a/central-asia-polygamy/29863234.html

"Why are men in Central Asia taking a second family?"

Kandiyoti, D. (1988, September), "Bargaining with Patriarchy," *Gender and Society* **2** (3), 274–290.

Korotkova, Elena (2017, 2 August), "Ia b imel trekh zhen: v Kirgizii snova govoriat o legalizatsii mnogoznenstva," accessed 9 September 2019 at http://www.news-asia.ru/view/10637

Короткова, Елена, "Я б имел трёх жен: в Киргизии снова говорят о легализации многоженства," 02.08.17, NEWS-ASIA: http://www.news-asia.ru/view/10637

"I would have three wives: In Kyrgyzstan they are again talking about the legalization of polygyny."

Kozhonaliev, S. K. (2000), *Obychnoe prava kyrgyzov*, Bishkek: Fond Soros-Kyrgyzstan.

С.К. Кожоналиев, *Обычное права кыргызов* (Фонд Сорос-Кыргызстан, Бишкек, 2000).

Customary Law of the Kyrgyz.

Kushner, P. (1929), *Gornaia Kirgiziia (Sotsiologicheskaia Razvedka)*, Moscow: Izdanie Kommunisticheskogo Universiteta Trudiashchikhsia Bostoka imeni I.V. Stalina.

П. Кушнер, *Горная Киргизия (Социологическая Разведка)* (Moscow: Издание Коммунистического Университета Трудящихся Востока имени И.В. Сталина, 1929).

Mountainous Kyrgyzia.

Momunov, Emilbek (2016, 22 September), "Komu dozvoleno brat' btoruiu zhenu?" accessed 9 September 2019 at https://www.gezitter.org/interviews/53622_komu_dozvoleno_brat_vtoruyu_jenu/

Эмилбек Момунов, "Кому дозволено брать вторую жену?" *Gezitter.org* 22.09.2016: https://www.gezitter.org/interviews/53622_komu_dozvoleno_brat_vtoruyu_jenu/

"Who is allowed to take a second wife?"

Nixon, Rob (2011), *Slow Violence and the Environmentalism of the Poor*, Cambridge, MA: Harvard University Press.

"'Odnazhdy nash muzh ele raznial nas.' Istorii vtorykh zhen" (2017, 1 December), *Vchernii* Bishkek, accessed 9 September 2019 at http://www.vb.kg/society/odnajdy-nash-myz-ele-raznial-nas-istorii-vtoryh-jen.html

"'Однажды наш муж еле разнял нас.' Истории вторых жен," *Вчерний Бишкек* 01.12.2017: http://www.vb.kg/society/odnajdy-nash-myz-ele-raznial-nas-istorii-vtoryh-jen.html

"'One day our husband barely pulled us apart.' A history of second wives."

Schatz, Edward (ed.) (2009), *Political Ethnography: What Immersion Contributes to the Study of Power*, Chicago: University of Chicago Press.

Tynaeva, Nurzada (2016, 15 March), "Den' dvoezhentsa – prazdnik na troikh," accessed 9 September 2019 at https://knews.kg/2016/03/15/den-dvoezhentsa-prazdnik-na-troih/

Нурзада Тынаева, "День двоеженца – праздник на троих." 15.03.2016. *KNEWS*: https://knews.kg/2016/03/15/den-dvoezhentsa-prazdnik-na-troih/

"The day of bigamy – a holiday for three."
Ugolovnyi Kodeks Kyrgyzskoi Respubliki (Bishkek: Akademiia, 2018).
Уголовный Кодекс Кыргызской Республики (Bishkek: Академия, 2018).
Criminal Code of the Kyrgyz Republic

11 When rednecks became meth heads: cultural violence, class anxiety, and the spatial imaginary

Aaron H. Gilbreath

Introduction

This chapter explores the way discourses of class, place, and methamphetamine use become intertwined to create an anonymous 'other' worthy of suffering the incremental yet calamitous effects of the constant denial of basic needs associated with slow violence (Davies 2018). Specifically, I trace the way tropes regarding methamphetamine users have been grafted on to rural whites as an ongoing act of cultural violence that facilitates their continued hardship. Rob Nixon (2011a) defines slow violence as harm that occurs 'gradually and out of sight, a violence of delayed destruction that is dispersed across time and space, an attritional violence that is not typically viewed as violence at all' (p. 2). Nixon's slow violence is temporally dispersed, meted out over decades or more on specific populations. More often than not, these populations are poor, lacking the political capital to defend themselves. These victims are 'the casualties most likely not to be seen, not to be counted' by the population at large (p. 13). Nixon's work emphasizes environmental degradation as a form of slow violence but allows for the existence of slow violence in myriad other forms (Nixon 2011b).

Nixon's work builds on Johan Galtung's (1969) theory of structural violence. Galtung's structural violence is the suffering created by the systemic denial of basic needs (Davies 2019). Under structural violence, economic and political systems create or enhance suffering through unequal dispersal of resources, rights, and justice. Nixon argues that slow violence differs from structural violence. While it shares a concern for social justice and imperceptibility,

173

Nixon is concerned more with the dimension of time, and how violence might endure long after the structures that made it possible have been removed. Slow violence has been a useful concept for geographers. Davies (2019) notes that as a spatial concept, it allows geographers to 'include the gradual deaths, destructions, and layered deposits of uneven social brutalities within the geographic here and now' (p. 2).

In his exploration of slow violence, Nixon foregrounds the need to be able to present and describe this violence in a manner that makes its slow-moving, latent effects urgent and apparent, as a means of contesting the largely invisible form of violence. Davies (2019) contests this imperceptibility of slow violence. He argues 'slow violence does not persist due to a lack of arresting stories about pollution, but because these stories do not *count* [author's emphasis]' (p. 3). Because the stories do not matter to society writ large, certain populations and landscapes are allowed to be harmed, experiencing Achille Mbembe's *necropolitics*, in which they suffer a long slow wounding, a 'death-in-life' (Mbebe 2003, as cited by Davies 2018, p. 1540). They are not necessarily actively harmed by authorities, rather they are allowed to suffer, degrade, and die slowly over time.

Galtung (1990) theorized how this suffering can be allowed to happen with his concept of cultural violence. Cultural violence is 'any aspect of culture that can be used to legitimize violence in its direct of structural form' (p. 291). It is the means by which structural or direct violence in Galtung's terminology, and slow violence in Nixon's, are allowed to occur. Cultural violence makes violence permissible by marking some members of the population deserving of their suffering. 'The culture preaches, teaches, admonishes, eggs on, and dulls us into seeing exploitation and/or repression as normal and natural, or into not seeing them (particularly not exploitation) at all' (p. 295). Invisibility does not remove effect. As Springer (2011) notes, the 'invisible geographies of violence' have very real, 'nonillusory' effects on human bodies (p. 92).

Cultural violence is enacted through discourse, what Wilson and Bauder (2001) define as 'a field of strategies (statements, views, theories, concepts, objects of analysis and their interrelations) that creates and perpetuates narrow veins of comprehension. It illuminates one way of seeing while simultaneously annihilating alternatives ... It creates realities that conceal the sameness of people and places' (p. 259). These discourses create distance between subjects, placing certain groups outside of society, making them unknowable and unworthy. One method of creating this intellectual and empathetic distance is through placing the perceived 'other' in a specific, imaginary landscape that helps to justify their social and class position.

Many geographers have studied the way in which abstract, imagined, or idealized spaces are deployed as a means of justifying social inequality and violence (Davies 2018, 2019; Elwood and Lawson 2013; Lawson et al. 2008, 2010; Springer 2011; Springer and Le Billon 2016; Wilson and Bauder 2001). Springer (2011) states that imaginary geographies 'fuse distance and difference together' to create an unknowable 'other' in a space that by necessity must be inherently different from the one we occupy (p. 93). Wilson and Bauder (2001) note that space is 'a richly used resource' in the discursive construction of others. 'It notably bonds subjects to imagined landscapes of cultural and moral illegitimacy, roots imagined landscapes of civility against which these people are compared, and makes these people transgressive as spatially violating beings' (p. 259).

Hegemonic imaginary geographies are accepted uncritically by the masses who view them as basic commonsense. Springer (2011) notes that our perceptions of places we have not visited are entirely dependent upon 'the imaginings that have been circulated, rendered, and internalized or rejected in forming our cartographic understandings' (p. 94). Furthermore, these cultural imaginings of place have real effects on our expectations for the people who occupy them. If one views a place as inherently violent, then they *naturally* expect that people residing in that place will experience violence. Elwood and Lawson (2013) have noted that spatial imaginaries are often deployed in times of economic crises as means of determining who gets to recover. Suburbs, for example, are not conceived of as places of poverty, so poverty in those sites must be remedied, while the inner-city and rural America are places where poverty exists unproblematically, and residents there need not participate in an economic recovery.

Imaginary rural white spaces have long done the work of creating an undeserving poor white subject in American culture. Because class is a constantly changing and contestable position, middle-class whites must distance themselves from their poor or working-class others, lest they fall into that status themselves. 'Middle-class anxieties stemming from rapid US economic restructuring are mitigated through representations of poor rural folks as "white trash", "criminal minorities", and "rednecks" who choose deviance' (Lawson et al. 2008, p. 738). Poor rural whites, whether working or not, are presented as lazy, ignorant, dangerous, and deviant, their poverty a lifestyle choice (Elwood et al., 2015). The process of making these subjects often relies on abstract rural spaces that are wild, underdeveloped, backward, and threatening (Jarosz and Lawson, 2002). The logic goes that the inhabitants of these places must be poor because the places are poor, and the places are poor because the people occupying them have been too backward, ignorant, or lazy to make proper use of the natural bounty presented to them.

Scott (2010) argues that these conceptions of rural space, and Appalachia in particular, have justified the creation of an environmental 'sacrificial zone' in the region. For Scott, residents of Appalachia are viewed as 'matter out of place' by American society. In the nineteenth century, their inability to transition from subsistence living and integrate into the national market economy represented a flaw inherent in the population. Today their class status in the face of whiteness that should have protected them from poverty marks them as lazy and undeserving. Society has decided that if they cannot or will not make proper economic use of their environment, then others should be allowed to exploit it, 'safe from the type of environmental disaster experienced by the poor' (p. 31). Thus, the spatial imagination of Appalachia and its residents does the cultural violence work necessary to allow the slow violence of coal mining and mountain-top removal.

These discursive parings of a marginalized other and an imaginary space enact cultural violence. Because cultural violence is the grease in the wheels of slow violence, there is merit in tracing how culturally violent discourse is created and applied. If cultural violence makes structural violence appear natural or right, then tracing the manner of its creation might help to shine light on the slow violence it enables. Many of the tropes mentioned above have largely existed since the nineteenth century (Scott 2010). However, in the last 40 years, a new stream of discourse has been woven into the cultural violence surrounding rural whites. Since the 1980s, methamphetamine use has become steadily conflated with white poverty and rurality. By tracing the manner in which this discourse is created, and the important role imaginary places play in the process, we can help to illuminate what Springer (2011) calls 'the *how* and the *where*' of slow violence (p. 91). In unravelling the origin of this trope, we'll see that discourses contributing to this particular cultural violence have been created by experts in medicine and the social sciences, policing agencies, news media, anti-drug campaigns, and popular television. We will begin with the near present, where the discourse is already fully formed to see how it functions today, and then move back to the beginning to see how it came to be this way.

The methiest states

In October of 2013, *The Huffington Post* published an article and infographic with the headline: 'The Methiest States in the US.' The graphic had a choropleth map of methamphetamine lab seizures by state for the Midwest in 2012 overlain on a background of green smoke familiar to anyone who watched the television series *Breaking Bad* (*Huffington Post* 2013).[1] The infographic itself had the title, 'The Midwest Breaks Bad.' During the methamphetamine epidemic of the mid-2000s, this type of article became a rite of passage for newspapers and websites hunting for page hits, appearing year after year after year, despite the fact that either the state of Missouri or Tennessee has remained number one in methamphetamine lab seizures since 2003, according to the National Clandestine Laboratory Seizure System, which began keeping such statistics in the year 2000 (Missouri State Highway Patrol 2019).

The level of fascination people have with methamphetamine use and users is entirely disproportionate to the drug's presence in society, and has been so since the 1960s (Jenkins 1999). At the time of the *Huffington Post* article, regular methamphetamine users made up just two-tenths of one percent of the population. Their presence was significantly lower than that of users of cocaine (.06 percent), hallucinogens (.4 percent), or heroin (.3 percent), yet we rarely if ever see articles written about the 'the cocaine capital of the United States' (Substance Abuse and Mental Health Services Administration 2014a).

Presumably, labeling places as methy means that they are places where the population is using methamphetamine at a high rate. Articles such as the one by the *Huffington Post*, as well as others that rush to identify the 'meth capital of the United States,' usually use meth lab seizures as their measure. Superficially, this makes some sense. Meth labs are the feature that makes methamphetamine a sexy geographical topic. Unlike cocaine or heroin, which are imported into the United States, methamphetamine can be made anywhere using just a few household chemicals and some cold medicine. These sites of production are called meth labs. Presumably, if there are a large number of meth labs found in an area, they must be there to meet an incredible amount of demand. However, there are a number of issues with this logic.

Since the passage of the Combat Methamphetamine Epidemic Act of 2006, which placed pseudoephedrine-containing pills behind the counter, and made their purchase more difficult, the overall size of seized meth labs has been shrinking, designed only to produce a very small amount of product for consumption by the cook and perhaps a few acquaintances. Between 2006 and 2011, 81 percent of labs seized by police produced less than two ounces of meth, and more often than not were the size of a two-liter soda bottle (National Drug Intelligence Center 2011).

In truth, methamphetamine lab seizures do not correlate particularly well with other measures of methamphetamine use. If we look at Treatment Episode Data Sets of people seeking admission for drug treatment who cite methamphetamine as their primary problem drug (which one might argue is far more indicative of meth use and, therefore, 'methiness,' than lab seizures), we see that California and Oregon both outranked Missouri in raw treatment mentions at the time of the *Huffington Post* article. If we calculate a rate per 100,000 people in the population, Missouri ranks tenth. The other states with extreme meth lab numbers in the *Huffington Post* article, Kentucky, Illinois, and Tennessee, all ranked in the twenties in terms of treatment admissions (Substance Abuse and Mental Health Services Administration 2014b). There was no correlation between number of meth labs seized in a state and the number of people in that state who sought treatment for methamphetamine abuse. Similarly, if one looks at methamphetamine mentions in the Metropolitan Statistical Areas included in the Drug Abuse Warning Network, we find that methamphetamine mentions are highest in Phoenix, Seattle, and San Francisco. Arizona and Washington, however, each had under ten labs seized in 2012, a far cry from the 1,825 labs seized in Missouri. The entire state of California, which has been the epicenter of methamphetamine use since the 1960s, had less than 80 labs seized (Substance Abuse and Mental Health Administration 2014c; Missouri State Highway Patrol 2019).

The lack of correlation between meth lab seizures and indicators of abuse stems from the fact that the vast majority of the methamphetamine consumed in the United States is not produced here at all. It is produced in large labs in Mexico, and then smuggled into the country by the Mexican drug-trafficking organizations that dominate most drug commodity chains in the Western hemisphere. In fact, domestic labs only account for about 20 percent of the methamphetamine supply. Large numbers of lab seizures simply represent areas where there is not a large enough market to attract imported methamphetamine distribution (Gilbreath 2015).

So, while it is interesting to track meth labs, and they can tell us a lot about where people are cooking meth here in the United States, they tell us very little about where people are using the drug, presumably what we mean when we use a term like 'methy.' However, I do not believe that was the goal of the *Huffington Post* infographic or any other article looking for the meth capital of the country. Calling a place 'methy' is simply a new code for conceiving of a place as poor, and its population as white trash. Meth seizures are emphasized precisely because they are high in rural states like Missouri and Tennessee, and confirm the already existing stereotypes for those places, as well as those for meth users. The dominant cultural tropes regarding rural places tell us that their populations are white, dirty, poor, lazy, ignorant, and prone to criminality. These same characteristics have been used since 1969 to describe methamphetamine users. These stereotypes have become so hegemonic that it makes perfect sense in the cultural imaginary for the populations of rural places to be using a lot of methamphetamine (Hartigan 2005; Jarosz and Lawson 2002; Lawson et al. 2010; Scott 2010).

How a wonder drug breaks bad

Methamphetamine users have occupied a low-status position in the nations' drug-using population since the late 1960s, but it was not always this way. When the amphetamine family of drugs was first introduced, they were hailed as wonder drugs with the potential to cure the many ailments of modern society. A 1936 headline in the *Chicago Daily Tribune* claimed the drug 'ends [the] urge to suicide' (*Chicago Daily Tribune* 1936). Dr. Abraham Myerson told the American Psychological Association that amphetamines had been 'found useful in the treatment of nervous disease caused by the swift tempo of modern living' (Cutter 1937). As the drugs increased in popularity around the country, an editorial in the *Annals of Internal Medicine* associated its use with the American Dream:

> It is possible that amphetamine popularity reflects American culture ... The amphetamine user is a caricature of many widely admired America traits: intense activity, efficiency, persistence and drive, and the desire to excel, to break records, and to move with ever-greater speed. (Edison 1971, p. 609)

The adulation was short-lived. The late 1960s saw methamphetamine (and all amphetamines) go from being associated with the American Dream to becoming the scapegoat for the end of the Summer of Love in San Francisco. San Francisco and other urban centers like the East Village of New York became the first locations in the country to see high-dosage methamphetamine injection. It was in San Francisco that the term 'meth head' first appeared in the 1960s to describe injectors of Methedrine, the Burroughs Welcome brand of prescription methamphetamine. Though meth head has been the term that has stuck, another term, 'speed freak,' was also used to describe the methamphetamine-injecting denizens of the counter-culture movement.

Several studies of methamphetamine users in the Bay Area from the 1960s attempted to classify who was using what among the counter-cultural movement. In these seminal studies, the Methedrine users were constructed in opposition to users of LSD (lysergic acid diethyamide, also commonly called 'acid'), a hallucinogen popular among hippies. In one account, users of hallucinogens were described as being found 'more often among the older, more established and less-transient members of the community' (Davis and Munoz 1968, p. 161). They used drugs 'for purposes of mind expansion, insight, and enhancement of personal attributes' (p. 160). Meth heads, on the other hand, described also as 'speed freaks' in this article, were in search of 'drug kicks.' They were found among the 'more anomic and transient elements of the community, in particular, those strata where "hipness" begins to shade off into such quasi-criminal and thrill-seeking conglomerates as the Hell's Angels and other motorcyclists' (p. 160). The authors also made class comparisons. Acid users were 'by and large persons of middle and upper middle class social origins whereas "freaks" were much more likely to be of working class background.' Finally, they concluded, 'LSD equals self-exploration/self-improvement equals middle class, while Methedrine equals body stimulation/release of aggressive impulses equals working class' (Davis and Munoz 1968, p. 161).

Speaking in 1969 to the staff conference in the Department of Medicine, at the University of California Medical Center, the head of the Haight-Ashbury Medical clinic, Dr. David E. Smith, lamented the fact that the 'speed freaks' had 'driven away' the acid heads of Haight-Ashbury (Smith 1969, p. 151). Dr. Smith evoked an idealized Haight-Ashbury community in his testimony, lamenting the harm that had come with meth users who had forced proper hippies to seek out a separate space for their higher aspirations:

> The "acid head" community cannot live with the "speed freak" community because of the violent characteristics of the latter. As a result the "hippies" have left the Haight-Ashbury district, moving to the country where they can establish small rural communes which tolerate and reinforce their belief systems. Unfortunately, in the conflict between "speed freaks" against "acid heads," speed always drives out acid just as in the broader society the philosophy of violence always dominates the higher aspirations of non-violence, peace, and love. (p. 156)

The idea that the meth heads ruined San Francisco for the hippies has stuck in the public consciousness. Thirty years later, writing in *Rolling Stone*, Peter Wilkinson states without any hesitation, 'meth arrived on the Haight-Ashbury scene in the late 1960s and ruined the Summer of Love vibe. Hippies mainlining speed – it was the wrong mix of pharmaceutical and subculture' (Wilkinson 1998, p. 51).

Not every user of methamphetamine was shooting up. One *New York Times* article estimated that across the country, 'thousands and probably millions use amphetamines without becoming wild-eyed "speed freaks." They drop pills to finish papers, wax floors, lose weight, write songs, sing songs, have conferences, sculpt, wake up and think more clearly' (Black 1970, p. 198). Even in the Bay area, and despite the lamentations of Dr. Smith, more level-headed analysis showed that the cohort of Methedrine users included 'ex-students, talented artists or writers, [who were] apparently equipped to be as successful as the average 20 year-old college student' (Carey and Mandel 1968, p. 171). A survey of research literature at the time found conflicting reports of exactly who made up the using cohort, reflecting a diversity not found in the problematic accounts of Smith or Davis and Munoz (Cox and Smart 1970).

Whatever the truth was, the classing of methamphetamine users begun in the 1960s stuck. Whatever class and ethnic diversity that existed in this original using cohort, which we know also included African Americans, factory workers, and members of the gay community, was erased from public consciousness, replaced with the image of the working-class methamphetamine user who was white, potentially violent, certainly had criminal intentions, and was looking for a body high (Jenkins 1999).

After the Controlled Substances Act of 1970, methamphetamine was driven underground. The little bit that was available on the streets was primarily produced and distributed by outlaw motorcycle gangs (Gilbreath 2014). The nickname 'crank' comes from the fact that bikers used to hide methamphetamine in the crankcases of their bikes to smuggle it (Owen 2007). That these gangs were largely white and wore many of the trappings of white supremacists helped cement the reputation of methamphetamine being a white person's drug. The gangs' criminal activities also helped reaffirm the notion that meth users were violent.

Meth heads as rednecks

When, in the late 1980s, methamphetamine production began to boom around San Diego, law enforcement there did their part to continue the trope of violence and methamphetamine. In doing so, they deployed a different spatial imaginary. These agents established their credibility on the subject of methamphetamine by dubbing their city 'the meth capital of the world' and comparing it to Bogotá (Wiedrich 1987). By raising the specter of 1980s Bogotá as a direct comparison to their city, DEA agents brought to mind the spatial imaginaries of South American lawlessness and violence that news coverage of the War on Drugs had projected onto American television sets throughout the decade. For DEA agents and administrators, the rhetorical move of comparing San Diego to Bogotá created a sense of urgency in the American public that could be used to increase funding for the agency and raise its standing within the governmental bureaucracy. The violent imaginary evoked by the claims of the local agents was given additional credibility in 1995, when a local man who was high on methamphetamine stole a tank and drove it through downtown San Diego (Jenkins 1999; Owen 2007).

While claiming credibility and importance by emphasizing the scope of the methamphetamine problem in San Diego, DEA agents simultaneously disparaged the significance of the drug and its users, referring to methamphetamine in the press and in presentations to Congress as 'redneck cocaine' (Jenkins 1999). A derogatory moniker, the term 'redneck' has long been used to police the line between middle and working-class whites (Hartigan 2005; Jarosz and Lawson 2002; Lawson et al. 2010; Scott 2010). The construction of the rural white subject as a 'redneck' or white trash does considerable work to hide the economic processes that have caused enduring rural poverty (white or otherwise) in the first place. Jarosz and Lawson (2002) identify three ways that 'redneck discourse' works to naturalize rural white poverty. The first is to present rural working-class whites as obsolete, clinging to professions that they should have given up long ago in the face of economic restructuring. The second is to define class as a lifestyle choice, where consumption patterns and cultural variables become class markers, rather than adaptations to the suffering created by poverty. Finally, white racism is reconstructed as 'redneck' racism, absolving middle- and upper-class whites of their own racist beliefs and practices. In defining white poverty as a choice reflected in things like profession and consumption choices, marking rural whites as obsolete and potentially criminal enables the slow violence of economic neglect and enduring rural poverty.

Grafting methamphetamine, which already had the reputation for being used by whites who were potentially violent onto the redneck stereotype instantly made the kind of natural sense one associates with cultural violence. In this light, the decision to use meth becomes just another cultural marker indicating 'poor' consumption choices. This type of name, made by presumed experts in a position of authority, allowed others in regions of the country where methamphetamine had yet to spread to assess the drug as a 'redneck' problem, confining its use to a specific cultural trope around 'white trash.' Rich whites who can afford it, use cocaine; poor whites who cannot, use meth instead.

It is not surprising then that the media of the time quickly and unapologetically adopted the terminology. In an analysis of media coverage of the 1990s, Jenkins (1999) found that the *San Francisco Chronicle, Christian Science Monitor, U.S News & World Report, New York Times,* and *National Public Radio* among others all used some variation of the term 'redneck cocaine' in breathless coverage of the drug's potential spread among poor whites. Most news stories emphasized that methamphetamine was moving away from the American West and into the Midwest. In these instances, one can see the rhetorical knots the media tied itself into when trying to couple the American Heartland with the poor, ignorant, and violent whites characterized by the redneck cocaine moniker. This internal stress is perhaps best exemplified by the 1998 article 'America's drug: Postcards from Tweakville' which appeared in *Rolling Stone* in February of 1998 (Wilkinson 1998). In the article, Wilkinson careens wildly from describing Independence, Missouri as a quintessential Midwestern town with a 'postcard perfect town square,' to painting the home of a specific meth user as 'a single-story dump surrounded by cars, washing machines and lawn mowers that had been disassembled and then put back together. A tweaker's place, with that Sanford and Son feel' (p. 50). The users' yard is deemed not to belong in a town with an 'aura of Midwestern rectitude.' It represents the corruption of the white middle-class utopia of the Midwest. This user's home and poverty are an example of Scott's (2010) 'matter out of place.' They are a threat that anyone (anyone white) can backslide from their position of privilege through methamphetamine. Methamphetamine had become 'a symbol of white degeneracy, of massive downward social mobility' (Jenkins 1999).

At the beginning of methamphetamine's re-emergence as a significant problem drug, the media's focus on its spread to the predominantly white suburbs and rural areas of the Midwest ignored large swaths of the drug's using population. The 'upscale gays, urban manual workers, college students, and professional athletes,' along with African and Asian Americans who also used methamphetamine were erased from the discussion (Jenkins 1999, p. 140). The discourse winnowed the user population to only poor whites, obliterating other potential avenues of investigation, intervention, and treatment.

In the early 2000s, several well-intended anti-meth campaigns served to further stigmatize methamphetamine use and reinforce the discursive connection between the rural white working class and methamphetamine. The Faces of Meth program was an informational presentation and slideshow created by a Multnomah County Sheriff's Deputy in 2004 to warn students away from methamphetamine use. The presentation shows mugshots taken at different points in time of the same person. The images are striking. Within the space of three months to two years each subject has experienced dramatic physical degradation. They are universally more wizened, haggard, and injured. They perfectly embody the 'living-death' created by slow violence (Davies 2019). The implication, true or not, is that the changes that these poor souls (all of whom are white) have undergone are entirely the result of their methamphetamine use. However, the goal of this presentation is not to engender empathy or support for these users. The goal is to scare future users, to warn them off ever trying meth.

While there is no evidence that these types of scare tactics work, the images, which are available for free on the Web, have been among the most repro-duced, shared, and meme-ified of the mid-2000s (Linnemann and Wall 2013; Linnemann 2016). The popularity of these images is not due to an outpouring of sympathy for the subjects pictured in them, but to the way they reinforce the tropes of the underserving white poor. As Linnemann and Wall (2013) explain, 'the "faces" are structured by and embedded within long-standing cultural-economic anxieties about the figure of "white trash" and the domi-nance and precariousness white social position' (p. 321). The pictures exist as a warning about class backsliding and as targets for derision.

The Meth Project produces similar images for public consumption. However, rather than use images of real users, the project creates fictionalized accounts of the type of class backsliding that can occur with meth use. One television ad depicts two adolescent girls prostituting themselves outside of a truck stop. As the two young girls take the men who will pay them for sex to a public restroom to consummate their transaction, the girls' appearance transforms. Their clothing and hair become disheveled and stained, their faces gaunt and covered in sores. The narration of the ad says, 'This isn't normal, but on meth it is.' Another print ad shows two teens beating and robbing an elderly man with the caption 'beating an old man for money isn't normal. But on meth it is' (Meth Project 2016). Prior to the release of Spanish Language Meth Project ads, every teen depicted in the spots was white and sported a cadaverous physique, open sores, and meth-mouth: the rotting teeth and ulcerated and cracked lips associated with extreme and prolonged methamphetamine abuse.

The Meth Project began in Montana as the Montana Meth Project, a campaign originally funded by billionaire Thomas Siebel, who feared what the drug might do to the predominantly white state where he resided part time. As is often the case with methamphetamine, the ads reflect a hysteria that was not merited by actual usage statistics. Meth use among teens in Montana had been in steady decline for seven years before the campaign began. Rather than reflect reality, the ads demonstrate the class anxiety inherent in the campaign's founding. Each image says more about fear of white poverty than it does the dangers of methamphetamine.

The imagery of Faces of Meth and the Meth Project have found their way into popular media. Linnemann (2010) notes that most media depictions of methamphetamine users use "overtly sexualized, racialized, and gendered" caricatures (p. 98). Examples abound in popular and prestige television. In the first season of HBO's *True Detective*, the primary suspect of a series of murders is described by one of the main characters as 'the Devil.' Much of the first half of the season is spent documenting how this character is evil incarnate. When he is finally revealed, he is a Southern man named Reggie Ledoux (Fukunaga 2014). Reggie represents the nexus of the white trash and meth head stereotypes. He is a meth cook who lives in a trailer in the rural South, is covered in Nazi tattoos, and has been kidnapping and raping small children. Reggie is dirty, violent, and racist because he is white trash, and because he is white trash, he also cooks methamphetamine. He is an amalgamation of white middle-class fear.

The meth user as white trash has also infiltrated sitcoms as an easy way to engender laughter. Perhaps the best example can be found in a season 7 episode of NBC's *30 Rock*. In that episode, Kenneth, the innocent former NBC page-turned-janitor from Kentucky is debating how to vote on a question of whether or not to empower the mayor of his hometown, Stone Mountain, Kentucky, to fix the town's clock tower. The show does a quick cutaway to the mayor herself, Debbie, who is extremely agitated. She is disheveled, her hair is a mess, she has scabs on her hands, and her teeth are yellow. Mayor Debbie explains, with increasing tension and anxiety, that they should let her fix the tower herself, as she wants 'to see how it works.' She goes on to say that her friend Jo-Jo did it with a toaster and it still works. She finishes by screaming, 'LET ME DO IT!' (Riggi 2012) Without saying it explicitly, Mayor Debbie is coded as a meth head. Her clothes and hair are a mess. She is gaunt. Her dry lips and rotten teeth are classic examples of meth-mouth. Her rising anxiety levels indicate a person who is sped up. Finally, she is from Appalachia. The logic is implicit. Because Debbie is from Appalachia, she must be a meth head, even if she is mayor of her town.

Conclusions

The construction of the meth-using subject has long utilized the same tropes as those around poor rural whites, mimicking the stereotypes of the redneck and white trash. This discursive creation has served the purpose of boundary making, drawing a line between what is acceptable white behavior and what is not (Wray 2006). Frequently this stereotype has relied upon imagined places in its construction. In some cases, as in the Midwest or San Francisco, white methamphetamine users have been blamed for the decline of an abstract white utopia. They represent the dangers of class decline, the precariousness of privilege. In other cases, the existence of methamphetamine use or production in a region has reinforced existing stereotypes in the discursively created public imagination of those places. These places naturally have meth labs because they are already dirty, and the people there use methamphetamine because they are poor and white. The pairing of methamphetamine and poor white spaces has become another way to blame the poor for being poor. In this sense, the coupling of place and drug has become an act of Galtung's (1990) cultural violence, reproducing discourses that enable the denial of treatment, support, and policies that might ameliorate the suffering produced by enduring poverty. These discourses also elide significant other groups in the methamphetamine-using community, limiting the possibilities for these other affected groups.

Note

1. *Breaking Bad* is an American television series about a teacher diagnosed with terminal cancer who takes up methamphetamine production to make sure his family will be provided for financially after he dies. It aired on the AMC network from 2008 to 2013.

References

Black, J. (1970), 'The speed that kills, or worse', *The New York Times*, 21 June 1970.

Carey, J. and J. Mandel (1968), 'A San Francisco Bay area "speed" scene', *Journal of Health and Social Behavior*, **9**(2): 164–174.

Chicago Daily Tribune (1936), 'Reports finding drug that ends urge to suicide: Claims discovery also cures nervous ills', *Chicago Daily Tribune*, 3 September 1936.

Cox, C. and R. Smart (1970), 'The nature and extent of speed use in North America', *Canadian Medical Association Journal*, **102**(7): 724–729.

Cutter, I. (1937), '"Blue Mondays" and Benzedrine', *Chicago Daily Tribune*, 21 June 1937.

Davies, T. (2018), 'Toxic space and slow time: Slow violence, necropolitics, and petrochemical pollution', *Annals of the American Association of Geographers*, **108**(6): 1537–1553.

Davies, T. (2019), 'Slow violence and toxic geographies: "Out of sight" to whom?', *Politics and Space C* (online but not yet printed) 1–19, accessed 12 August 2019 at https://journals.sagepub.com/doi/pdf/10.1177/2399654419841063

Davis, F. and L. Munoz (1968), 'Patterns and meanings of drug use among hippies', *Journal of Health and Social Behavior*, **9**(2): 156–164.

Edison, G. (1971), 'Amphetamines: A dangerous illusion', *Annals of Internal Medicine*, **74**(4): 605–610.

Elwood, S. and V. Lawson (2013), 'Whose crisis? Spatial imaginaries of class, poverty, and vulnerability', *Environment and Planning A*, **45**: 103–108.

Elwood, S., V. Lawson, and S. Nowak (2015), "Middle-class poverty politics: Making place, making people', *Annals of the Association of American Geographers*, **105**(1): 123–143.

Fukunaga, C. J. (2014), *True Detective*, season 1, episode 5, 'The secret fate of all life', aired 16 February 2014 on HBO.

Galtung, J. (1969), 'Violence, peace, and peace research', *Journal of Peace Research*, **6**(3): 167–191.

Galtung, J. (1990), 'Cultural violence', *Journal of Peace Research*, **27**(3): 291–305.

Gilbreath, A. (2014), 'West coast booms and east coast busts: Methamphetamine commodity chains of the 1970s and 1980s', *Historical Geography* **42**: 260–275.

Gilbreath, A. (2015), 'From soda bottles to super labs: An analysis of North America's dual methamphetamine production networks', *Geographical Review* **105**(4): 511–527.

Hartigan, John (2005), *Odd Tribes: Toward a Cultural Analysis of White People*, London, UK and Durham, NC: Duke University Press.

Huffington Post (2013), 'The methiest states in America', accessed 25 October 2013 at https://www.huffpost.com/entry/methiest-states

Jarosz, L. and V. Lawson (2002), "'Sophisticated people versus rednecks": Economic restructuring and class difference in America's west', *Antipode*, **34**(1): 8–27.

Jenkins, Philip (1999), *Synthetic Panics: The Symbolic Politics of Designer Drugs*, New York: New York University Press.

Lawson, V., L. Jarosz, and A. Bonds (2008), 'Building economies from the bottom up: (Mis)representations of poverty in the rural American Northwest', *Social & Cultural Geography*, **9**(7): 737–753.

Lawson, V., L. Jarosz, and A. Bonds (2010), 'Articulations of place, poverty, and race: Dumping grounds and unseen grounds in the rural American northwest', *Annals of the Association of American Geographers*, **100**(3): 655–677.

Linnemann, T. (2010), "Mad men, meth moms, moral panic: Gendering meth crimes in the Midwest', *Critical Criminology*, **18**(2): 95–110.

Linnemann, Travis (2016), *Meth Wars: Police, Media, Power*, New York: New York University Press.

Linnemann, T. and T. Wall (2013), "'This is your face on meth": The punitive spectacle of "white trash" in the rural war on drugs', *Theoretical Criminology*, **17**(3): 315–334.

Mbembe, A. (2003), 'Necropolitics', *Public Culture*, **15**(1): 11–40.

Meth Project (2016), *View Ads*, accessed 22 September 2019 at http://methproject.org/ads/tv/

Missouri State Highway Patrol (2019), *Meth Lab Statistics Disclaimer*, accessed 22 September 2019 at https://www.mshp.dps.missouri.gov/MSHPWeb/DevelopersPages/DDCC/methLabDisclaimer.html#

National Drug Intelligence Center (2011), *National Drug Threat Assessment: 2011*, Johnstown, PA: Department of Justice.

Nixon, Rob (2011a), *Slow Violence and the Environmentalism of the Poor*, Cambridge, MA: Harvard University Press.

Nixon, R. (2011b), 'Slow violence', *Chronicle of Higher Education*, **57**(40): B10–B13.

Owen, Frank (2007), *No Speed Limit: The Highs and Lows of Meth*, New York: St. Martin's Press.

Riggi, J. (2012), *30 Rock*, season 7, episode 5, 'There's no I in America', aired 31 October 2012 on NBC.

Scott, Rebecca (2010), *Removing Mountains: Extracting Nature and Identity in the Appalachian Coalfields*, Missoula: University of Montana Press.

Smith, D. (1969), 'Changing drug patterns in the Haight-Ashbury', *California Medicine* **110**(2): 151–157.

Springer, S. (2011), 'Violence sits in places? Cultural practice, neoliberal rationalism, and virulent imaginative geographies', *Political Geography*, **30**, 90–98.

Springer, S. and P. Le Billon (2016), 'Violence and space: An introduction to the geographies of violence', *Political Geography*, **52**, 1–3.

Substance Abuse and Mental Health Services Administration (2014a), *Results from the 2013 National Survey on Drug Use and Health: Summary of National Findings*, Rockville, MD: Department of Health and Human Services.

Substance Abuse and Mental Health Services Administration (2014b), *Treatment Episode Data Set (TEDS) 2002–2012*, Rockville, MD: Department of Health and Human Services.

Substance Abuse and Mental Health Services Administration (2014c), *The DAWN Report: Emergency Department Visits Involving Methamphetamine: 2007 to 2011*, Rockville, MD: Department of Health and Human Services.

Wiedrich, B. (1987), 'Dope merchants make San Diego capital of "speed"', *Chicago Tribune*, 20 April 1987.

Wilkinson, P. (1998), 'America's drug: Postcards from Tweakville', *Rolling Stone*, **780**: 49–55.

Wilson, D. and H. Bauder (2001), 'Discourse and the making of marginalized people', *Tijdschrift voor Economishe en Social Geografie*, **92**(3): 259–260.

Wray, Matthew (2006), *Not Quite White: White Trash and the Boundary of Whiteness*, London and Durham, NC: Duke University Press.

12 The slow violence of law and order: governing through crime

Samuel Henkin and Kelly Overstreet

Introduction

Geographic scholarship on security and violence tends to focus on 'the spectacle'—momentary aberrations of violence (Gregory and Pred 2007; Springer and Le Billon 2016). Violence cannot, and should not, be divorced from its material existence and its embodied experiences and accounts (Buzan 1991; Wyn Jones 1999). Yet, principally focusing on visible spectacles of violence displaces other forms of violence, including slow violence. The intensity and immediacy of visible forms of violence renders violence that occurs gradually over time and out of sight invisible (Nixon 2011, 2). Unsettling the dominant, yet equivocal, imaginaries of spectacular violence becomes necessary to see slow violence. In other words, to see slow violence, we must locate it. Locating slow violence in urban governance entails identifying displaced experiences and information, as well as looking for alternative perspectives rendered silent by dominant perspectives of crime and criminality in the *law and order* movement. Popularized as a conservative theme in the 1960s, law and order broadly refers to a more robust and strict criminal justice system organized and managed in an effort to control and prevent crime (Henson and Reyns 2015). We critically engage the concept of so-called 'law and order' as it continues to be, and historically was, exercised as a critical intervention in urban space that produces, enables and gives meaning to disciplining bodies and spaces, often violently.

While imagery of spectacular scenes of state violence, like police interventions, unfold in U.S. cities, the often-invisible role of policy decisions, legal systems and political or social response practices that serve to exert the state's monopoly of violence are often unnoticed or unacknowledged as violence. These administrative apparatuses and systems go unnoticed under pretenses of an apparent 'neutrality,' as the state evokes dominant narratives of safety and crime (McDonald 2019). 'Law and order' manifests through the power of discourse (narrative, materiality, embodiment and practice), shaping actions and consequences (O'Lear 2018). As such, 'law and order' embodies various discourses, materialities and practices that shape life chances, unevenly across space (Spade 2015). Adopting Nixon's (2011) definition of slow violence as a violence that is barely visible, dispersed across time and space and often not recognized as violence at all, we argue that slow violence is present in the day-to-day 'making' and 'doing' of governing through the maintenance of 'law and order.' As everyday urban politics and governance are increasingly understood through the lens of 'law and order,' it becomes consequential to consider the ways slow violence manifests: How does slow violence materialize in urban space? What are lived, everyday experiences of slow violence in urban space? In what ways does slow violence manifest through urban governance? Such questions pose new challenges in understanding the social and spatial relations of everyday life in cities, yet these questions tend to remain unasked, unanswered and unproblematized.

In this chapter, we call on geographers and other scholars alike to (re)examine through the lens of slow violence the everyday administrative decisions made and actions taken to govern urban space. Across U.S. cities, the state employs 'law and order' to advance a security governance arranged around the identifi- cation and discipline of bodies and spaces as criminal perpetuating unnecessary appeals to discourses of criminality. Punitive 'law and order' visibly resonates across U.S. political culture (Beckett 1997). Popular sentiments about crime continue to preserve myths of criminality and punishment despite decades of research and political reform on these issues. The adaptability and agility of 'law and order' to make sense of security 'realities,' rooted in public fear of crime, stabilizes the practice of 'governing through crime' (Simon 2007). 'Law and order' conjures fear of crime and victimhood to center its legitimacy in lawmaking and administrative practice, albeit unevenly, disproportionately impacting low-income communities, people of color and Indigenous and LGBTQIA[1] peoples across the U.S.

Our aim, then, is to demonstrate how the slower violence of 'law and order' might be made more visible and to begin to consider what types of indicators, measures and practices can be used to see and understand slow violence. We do so by considering the ways slow violence is innately embedded in our gov- erning administrative systems that are often complicit in maintaining relations of violence. We expressly focus on the U.S. legal system, highlighting the often-invisible roles of prosecutorial decision-making in and around crime. Through an analysis of policy and procedural documents, judicial decisions and media materials we expose slow violence in a series of seemingly innocuous everyday prosecutorial decision-making. Highlighting the often-invisible role of prosecutorial discretion brings to light the slow violence of prosecutorial decision-making from the decision to bring charges (and which charges) and dictating plea bargains to seeking life in prison, and even the death penalty.

Prosecutorial discretion is perceived to be value neutral in the justice system; however, the decision-making power of prosecutors is vast and largely unchecked (Spohn 2018). Court Justice Robert Jackson argued that 'the prosecutor has more control over life, liberty and reputation than any other person in America' (Jackson 1940, 3). We suggest that slow violence is situated at the intersection of governing through crime, whereby bodies and spaces are identified as criminal, and the increasing role of prosecutorial power in the service of urban governance. To advance our case, we locate slow violence in the legal proceedings of *The People of the State of New York against Korey Wise, Kevin Richardson, Antron McCray, Yusef Salaam and Raymond Santana*, also known as the 'Central Park Five,' whereby prosecutorial discretion resulted in gross negligence and violent injustice (Ghandnoosh 2014).

The slow violence of governing through crime

Governing through crime configures urban space in ways that have significant outcomes for spatial imaginaries of 'law and order' and (in)justice. As Delaney (2016) argues, 'Conventional spatial imaginaries tend to invisibilize injustices, obscure their contingencies and causes and uncouple injustice from responsibility' (268). The slow violence of (in)justice considers the ways in which law—its making and implementation—and violence intersect. In order to study, analyze and draw attention to the challenging questions of seeing and understanding the slow violence of law we turn to critical legal geographies and law and society scholarship (Reiz et al. 2018). This approach synthesizes streams of scholarship that make interconnections between law, spatiality and governance. Such an approach begins to untangle the ways law and legal administrative processes and practices contribute to slow violence. We think through the various ways the slow violence of law is socially and spatially selective in identifying and analyzing asymmetries of power and injustice (Blomley and Bakan 1992; Braverman et al. 2014).

We argue that the centerpiece of 'law and order' is fear of crime. Since the 1960s, fear of crime has been invoked as one of the most politicized issues in American society (Walker 2006). For over 50 years, research on fear of crime has attempted to understand and interpret perceptions of and experiences with criminal realities and victimization potentials (Ferraro 1995; Miller 2016). Inevitably there is no agreed-upon definition of fear of crime, its causes, factors or roles in society. Nevertheless, there is general agreement that fear of crime is emotive and affective, psychological, material and embodied (Zhao et al. 2015). Fear of crime experienced in the U.S. continues to remain stable over the past five decades, in spite of dramatic falls in reported crimes of victimization facilitating the expansion of 'law and order' across the U.S. (Grawert and Kimble 2019).

Fear of crime is shaped and nurtured by 'law and order' producing dominant security narratives of urban space as 'crime ridden,' 'dangerous' and 'unsafe' with social and spatial consequences. Urban space has long held perceptions of fear and insecurity that affect mobility (social and physical) as American cities change over time. Over-policing of urban communities of color is of significant concern due to the racialization of crime and criminalization of race which inflate crime rates in urban areas (Ward 2014). Fear of crime is innately connected to the social meanings of urban space and the perceptions of and attitudes toward crime associated with urban space (Chataway et al. 2017). As such, crime and victimization affects everyday geographies and interactions with urban space bound to different notions of safety (England and Simon 2010). 'Law and order' is power-laden in perpetuating fear discourses that define and maintain conceived boundaries of threat and safety, order and disorder and belonging and deviancy in urban space. These boundaries trap predominantly urban communities of color in racist spatial imaginaries and political mobilizations aimed to devalue their bodies and delegitimize their lived experiences of social, economic and political marginalization (Schuermans et al. 2010). Understanding how urban space is governed through crime makes delayed, dispersed and hidden slow violence visible.

The success of 'law and order' stems from its capability to exploit social and political fears of crime and victimization following historic patterns of injustice and criminalization of the 'other' (Hinton 2018). From the extermination of Indigenous peoples during colonial settlement to the onslaught of new criminal laws and penal systems during Reconstruction (e.g. Black Codes), the systematic criminalization of the 'other' legitimizes state violence and social control in the name of security (Ward 2014). With the rise of 'law and order' in the U.S., crime policy mobilized through conservative political backlash, or 'frontlash,' to the social and cultural upheavals of the 1960s brought about by the civil rights movement as well as the anti-Vietnam War, LGBTQIA and environmental movements (Weaver 2007). Furthermore, 'law and order' was militarized under the pretenses of a 'War on Crime' (also read as a war on poverty) stabilizing security narratives of urban spaces and residents as 'threats' to safety (Alexander 2012). Uncompromising penalties and punishment facilitated widespread political security reform in America erecting an enormous carceral complex. That carceral complex has repackaged the criminal 'other' and (re)defined U.S. inequality (Hinton 2018).

While most scholars point to the 1960s as the starting point to investigate 'law and order,' we understand 'law and order' as conditioned by a long trajectory of socio-political and historical forces—colonization, capitalism, racism (to name a few)—that constitute the circumstances for which the productive capacities of slow violence and criminality are spatially understood and unevenly applied. Determining and acting upon conceptions of criminality reproduces racist, classist, gendered, sexist, homophobic, transphobic, ableist, ageist and geopolitical discrimination and (slow) violence. We must acknowledge the historical legacies of these quotidian lived experiences. In doing so, slow violence becomes more visible in the ways governing through crime reflect, and effect, uneven destructive and neglectful impacts of 'law and order' that leads to unequal and unjust social and spatial relations.

Today, there are over 2.2 million citizens incarcerated in the U.S., 4.6 million people entangled in the carceral complex outside of prison walls (on probation or parole) and over 70 million people with criminal records (Grawert and Kimble 2019; Bazelon 2019). The legacies of the war on crime and its current manifestation of mass incarceration undoubtedly transform the logics of crime and security with significant socio-political consequences in urban space. There are numerous injustices of mass incarceration. The widespread incapacitation of people in the carceral complex is often caused by the methodical targeting of marginalized and disenfranchised peoples (Pettit and Gutierrez 2018). Recent scholarship on mass incarceration importantly interrogates the systematic and systemic violences of the very visible results of punishment in governing through crime (Alexander 2012). However, governing through crime is not solely about the act of punishment of those charged with criminal acts; in fact, governing through crime involves (re)producing subjects and social and political values through which to assert power in the pursuit of 'justice' (Simon 2007). In other words, governing through crime is often less about seeking justice and more about concentrating and retaining power from those who benefit from a system that creates and sustains inequities across space and society.

The practices of governing through crime stabilize a set of assumptions about crime and punishment that forms the consensus-building force of urban governance today. Yet, too often, the visible injustice and violence of order (policing, incarceration, post-incarceration reintegration) hide the nefarious slow violence of consensus-building and decision-making in our legal systems. An interrogation of the slow violence of law has yet to be fully realized. This process requires lively, creative and alternative perspectives on law and its capacity to produce uneven geographies of violence (Braverman et al. 2014). Seeing slow violence in our legal system necessitates engagement with the particularities of power. While law enforcement officers exercise a legitimate right to exercise violence (including the right to kill) and judges rule on questions of law, we consider the power of the prosecutor as most significant in perpetuating the slow violence of 'law and order.' As Sklansky (2016) contends, 'Much of what is wrong with American criminal justice—its racial inequality, its excessive severity, its propensity for error—is increasingly blamed on prosecutors' (474). Understandings of, and responses to, the rise of prosecutorial power are various, and at times competing (Sklansky 2016). However, in our ensuing analysis we privilege the significance of prosecutorial discretion to expose slow violence in urban governance. We offer a critical starting point for understanding how slow violence becomes visible in the everyday, seemingly banal decision-making of prosecutors governing through crime.

Seeing slow violence in prosecutorial discretion

Over 50 years ago, President Johnson's U.S. President's Commission on Law Enforcement and Administration of Justice (1967) recognized the influential role of prosecutors in the criminal justice system but condemned the highly discretionary decision-making practices of prosecutors. The Commission noted that prosecutorial decision-making often transpired 'hastily and haphazardly' with little to no oversight or review (130). The Commission was especially concerned with the informal, highly discretionary and mostly invisible processes surrounding charging and plea bargaining (Spohn 2018). The Commission called for more transparency and accountability and concluded that 'the greatest need is the need to know' (quoted in Spohn 2018, 323). We contend the 'need to know' is still the greatest obstacle in understanding and responding to prosecutorial discretion, and in particular, the slow violence it espouses.

Every decision maker in the U.S. carceral complex maintains forms of unchecked discretionary power that shape life chances, but prosecutorial discretion is often the most obscured (Tonry 2012; Sklansky 2016; Spohn 2018). Prosecutorial discretion continues to be shrouded in an epistemological and methodological fog. Unlike other agents and institutions of the US carceral complex, there is a lack of qualitative and quantitative data in prosecutorial offices (Barkow 2010; Tonry 2012). Moreover, prosecutors rarely disclose their decision-making process or even if a decision has been made. This lack of transparency is compounded by the 'boundary-blurring' performed by prosecutors (Sklansky 2016, 506). Prosecutors serve a mediating role between the order of policing and the law of the courts with no standard approach to exercising prosecutorial discretion. As such, the magnitude of prosecutorial discretion is fluid and transgresses all key aspects of governing through crime.

It is generally understood that exercising discretionary application of criminal law is the core of prosecutorial power (Wright and Miller 2010). Put simply, prosecutors make discretionary decisions to convict alleged criminals of alleged crimes. Single-handedly they get to decide which crimes to prosecute, whom to recognize as 'criminal' and thus charge, whether to offer a plea bargain, present concessions, divert a case or not, the level of aggression in pursuing a conviction, and what sentence to propose (Wright and Miller 2010). While prosecutors must act within the bounds of law, the legal limits placed on prosecutors is not extensive nor very restrictive. As Green (2019) contends, 'prosecutors generally decide for themselves both the processes by which they make discretionary decisions and the grounds on which they will make them' (597). Consequently, the decision-making impacts of prosecutors cannot be overestimated, not only for everyday individuals caught up in the carceral complex, but also in shaping the wider carceral complex itself. This point is especially true for marginalized urban communities, since prosecutors maintain significant power in shaping the urban fabric of American cities.

Prosecutors routinely deflect criticisms of their 'virtually absolute power' in deciding the life chances of countless individuals in the carceral complex by invoking the conceived neutrality of administrative systems of law (Miller 2004, 1252). By constitutional design, US courts are conceived to serve as neutral arbiters enforcing the state and citizen's obligations in the democratic law-making calculus committed to impartiality in the application of law; the so called 'neutral principles' (Wechsler 1959). Yet, the expectation of neutrality is rarely met (Kahan 2011). An ostensibly neutral legal system contributes to a depoliticization of law. By perpetuating the idea of apolitical realities of law, actors and institutions do not have to reconcile the ways socio-political and cultural meanings influence and interact with perceptions of law within the decision-making process (Kahan 2011). These claims to being an apolitical practice subtly erase the historical legacies of highly discriminatory law making, implementation and enforcement. Moreover, it is an insidious practice that denies the inherent violence systematically experienced within and as an effect of the U.S. legal system.

Prosecutorial discretion, cloaked in conceptions of neutrality and impartiality, depoliticizes the role of decision-making involving highly contentious and politicized phenomena that radically impacts the distribution of life chances (Spade 2015). Dismantling imaginaries of prosecutorial discretion as neutral and impartial becomes a means to make slow violence visible. Prosecutorial discretion and decision-making capacities of prosecutors must be interrogated in their full complexity to understand how 'law and order' produces uneven geographies of (in)justice and (in)security. As a starting point, we choose to highlight a critical stage of the decision-making process we consider to be most impactful in causing and perpetuating slow violence: *charging*. We grapple with prosecutorial discretion in charging in light of the ongoing 'innocence revolution' drawing attention to cases of wrongful imprisonment and exoneration (Marshall 2003; Medwed 2010; Zalman and Grunewald 2015).

Prosecutorial decision-making on charging

Of all the decisions made by the prosecutor, the initial decision to charge a suspect with a crime or not sets into motion a series of events that can forever alter the life chances of an individual (Medwed 2010). Prosecutors exercise unrestrained power in determining complex and controversial evaluations of proof and culpability and in the process govern remarkable decisions about an individual's character, liberty, and even life itself (Gershman 1993). Without losing sight of the affective and embodied impacts of this decision, it should also be noted that charging decisions shape, and are shaped by, the greater socio-political dynamics of 'law and order.' It is this initial decision to charge that prosecutors are particularly encouraged by 'law and order's demand for tough on crime practices' (Kohler-Hausmann 2018).

In 1978 the Supreme Court ruled that, 'so long as the prosecutor has probable cause to believe that the accused committed an offense defined by statute, the decision whether or not to prosecute, and what charges to file or bring before a grand jury rests entirely in his discretion' (*Bordenkircher* v. *Hayes*, 1978). Thus, it is generally understood that filing a criminal charge is dependent upon probable cause to believe in the person's guilt as understood by prosecutors. This is an inherently minimal threshold bound to highly subjective decision-making of the prosecutor regarding criminality and law. While various doctrinal rules, model ethical codes and state-specific rules offer a degree of guidance in regards to decision-making on charging, the sole discretion to charge lies with the prosecutor (Gershman 1993). The lack of transparency of this decision-making process begs the question: How do prosecutors arrive at a decision to charge especially as variance in charging results in significant life consequences?

Scholars working on prosecutorial discretion generally point to four motivations in charging decision-making process: legal, experiential, ethical and political (Gershman 1993). However, this motivation typology is an idealized version that lacks nuance and stabilizes the conceived neutrality of prosecutorial decision-making. Various studies examine the ways prosecutors exercise discretion in rejecting a substantial portion of cases at initial screenings (Spohn et al. 2014) and navigating questions of uncertainty (Albonetti 1987) while highlighting the importance of legal factors. Other studies expose evidence that the decision to charge reflects legally irrelevant subjectivities of the suspect and victim, like racial or ethnic identities (Smith and Levinson 2011; Stolzenberg et al. 2013). Additionally, the more amorphous aspects impacting prosecutorial decision-making like ambition, public sentiment, conviction mentality and personal bias/prejudice influence the charging process (Pfaff 2017).

The wide variance in charging decision-making significantly impacts how criminal law is applied across the U.S. and advances an unequal and uneven application of law. The foremost crisis of which is evidenced through mass criminalization and resulting social exclusion of predominantly nonwhite people (Ward 2014). While there is evidence that nonwhite people are not treated as well as whites—both as victims and criminal defendants—in every step of the criminal process prosecutorial charging has 'inextricable and profound' impact on racial disparities in the carceral complex (Davis 1998, 16–17; Taylor et al. 2018). For example, once charged, African Americans are 5.1 times more likely than whites to be incarcerated, and African Americans make up 38 percent of all U.S. prisoners although only representing 13 percent of the population (Glaze and Haberman 2013; Hetey and Eberhardt 2018).

The slow violence of racial disparities in the carceral complex starts well before the prosecutor's office. Continuing through the exercise of prosecutorial discretion in charging, the perpetuation of racial disparities and systematic malign neglect disproportionality fall on nonwhite people with lasting impacts even after time served. For instance, a collateral consequences report from the American Bar Association (ABA) identifies almost 45,000 U.S. laws that impede formerly incarcerated people from accessing a range of crucial services like educational assistance, housing and occupational licenses (Taylor et al.

2018, 158). It is vital to interrogate the slow violence engendered through prosecutorial discretion in charging throughout the entirety of the carceral complex.

An obvious question with respect to charging is: What if the prosecutor is wrong to charge? Growing interest in wrongful convictions demonstrates the dominant causal role of prosecutors in intentionally and unintentionally pursuing ill-advised charging decisions leading to wrongful convictions and imprisonment (Spohn 2018; Roach 2019). As Green and Yaroshefsky (2016) declare, 'there has been increased acceptance of the argument that prosecutorial misconduct is widespread and systemic' (52). The matter of wrongful convictions and the role of prosecutors in producing wrongful convictions is a critical juncture for geographers to theoretically, methodologically and practically engage slow violence. In the U.S., 'Sixty percent of all exonerations arose in just nine jurisdictions,' and eight of those jurisdictions are urban, exposing the extraordinary geographic inconsistency of wrongful convictions and exonerations (MacLean et al. 2015, 158). More troubling is that the majority of urban jurisdictions like the District of Columbia (DC), Chicago, New Orleans, Houston, and New York City (NYC) exceed national average exonerations rate. This uneven spatial distribution of exonerations requires more work to better understand how the application of 'law and order' and prosecutorial discretion disproportionately impacts people in urban places.

The National Registry of Exonerations has cataloged 2,507 exonerations over the last 30 years (The National Registry 2019). On average, exonerees serve almost 10 years of their sentence subjected to the brutalities of the carceral complex (The National Registry 2019). Exonerations are often misguidedly cited as evidence that the carceral complex works, in that exonerations 'resolve' a mistake (Webster 2019). This misconception belies the exceptional endeavor in achieving exoneration and the violence experienced throughout the process of doing so. We contend that exoneration cases are flashes of slow violence made visible that signify systemic violence of the legal system, but also serve as access points to more deeply investigate other cases of prosecutorial discretion that are less evident. As such, we offer a brief examination of ex-Manhattan 'law and order' prosecutor Linda Fairstein's exercise of prosecutorial power as a starting point for scholars to take up the call for a greater interrogation of everyday administrative policies and practices in governing through crime (Fischer 2019). Fairstein famously wielded prosecutorial discretion to orchestrate the wrongful conviction and imprisonment of the Central Park Five.

The Central Park Five is the moniker for the wrongful conviction and imprisonment of six young African American boys in NYC between 1989–1991.

The case, (re)popularized by the recent Netflix series *When They See Us* (2019), exposed the slow violence, collective failure and racial disparities of the carceral complex, and in particular the prosecutor's role in conviction. Fairstein charged the defendants with the brutal assault and rape of Trisha Meili in Central Park in 1989. Given the notoriety of this case, many details are well known (Color of Change 2019). Charges hinged solely on (false) confessions obtained the night of the assault. Fairstein's legal strategy relied on these confessions. Throughout seven hours of interrogation, police employed various problematic coercive interrogation techniques, during which no attorneys were present (Color of Change 2019). Despite the deceptive interrogation process, Fairstein decided to aggressively push the inclusion of confessions at trial even though every (false) confession had errors. Additionally, DNA and physical evidence found at the scene did not match any defendant. All but one defendant withdrew their confessions before trial and pled not guilty after legal representation realized their confessions were coerced. The presiding judge, Judge Thomas B. Galligan, admitted these confessions into evidence at Fairstein's behest. Fairstein exercised her discretionary power to dramatically impact the lives of five young boys, as well as Meili's healing process forever.

In the end, four defendants served 6–7 years in juvenile facilities (before transfer to adult facilities at age 21). Korey Wise, tried and sentenced as an adult at age 16, served almost 13 years in several prisons, including Rikers Island, before exoneration. In 2002, all convictions were vacated after convicted serial rapist Matias Reyes confessed to the crime, which DNA evidence confirmed. NYC reached a settlement with the Central Park five in 2014 for US$41 million; however, notably there was no admission of wrongdoing (New York City Press Office 2014). While Fairstein refuses to recognize any wrongdoing, a nuanced look into her career uncovers a legacy of slow violence in the exercise of her prosecutorial power. Throughout her career, Fairstein exercised prosecutorial discretion to advance policies and practices of 'law and order' (e.g. broken windows and zero tolerance policing) that disproportionately criminalized people of color, particularly cisgender and transwomen of color (Fischer 2019).

For example, Fairstein exercised prosecutorial power to enforce a highly contentious NYC loitering law that significantly profiled alleged sex workers. In doing so, she advocated for the protection of women by aggressively pursuing harsher punishment for the very same women who are uniquely vulnerable to violence: 'Fairstein contributed a distinct feminist justification for the mass policing and incarceration of women of color, who currently make up the fastest-growing prison population' (Fischer 2019). This 'carceral feminism' perpetuates flawed conceptions of increased policing, prosecution and imprisonment as the solution to violence against women (Law 2014). It fails to address the ways race, class, sex and gender identity shape vulnerabilities and violence, and it overlooks how greater criminalization places these same women at risk. Additionally, this form of 'carceral feminism' does not acknowledge violence experienced within the carceral complex. Our challenge is not only to locate slow violence in governing through crime, but to leverage our knowledge to provide meaningful interventions in the pursuit of an equitable system of justice; one that does not compound the inheritance of injustice that 'law and order' safeguards.

We believe exoneration cases are important initial entry points in assessing the hidden mechanisms that perpetuate incidents of slow violence. However, further exploration into prosecutorial discretion throughout the criminal justice process is required. Furthermore, future research on prosecutorial discretion should be further situated to assess both the systemic and local conditions that shape slow violence within urban governance. We invite scholars to consider the invisible mechanisms of urban governance that may occur at the courthouse in their own town or city A first step in this process is for researchers to explore legal cases, legal records and media coverage of district attorneys' offices in their own communities.

Conclusion

Across American cities, the unquestioned practices associated with a 'law and order' approach arrange people and spaces through criminality and (re) produce uneven levels of vulnerability to exploitation, oppression and violence. Our focus is to bring to light the hidden or unacknowledged violences of the often-invisible role of law and decisions in the legal systems that serve urban governance. If we understand slow violence as ever present, informed by historical legacies of imprisonment and criminalization, how do we best investigate this kind of violence? How do we best understand and respond to the unremitting, yet often concealed, forces operating at multiple sites and spaces of law and administrative decision-making that underpin slow violence? Working toward recognizing, and ultimately resolving, the injustices of slow violence of 'law and order' requires us to consider alternative spatial imaginaries. It requires a critical politics that deeply considers what law is, what kinds of violence and power emanate through law and its actors, how the social and spatial selectivity of law fundamentally impacts the distribution of life chances, and how laws and accompanying decision-making can harm people. In other words, we understand slow violence as a permanent feature of law, order and crime governance as we know it. It is this starting point where we ask geographers and other scholars to make their intervention in imagining a more just future by exposing the slow violence of 'law and order.'

Note

1. LGBTQIA is a contested umbrella intitialism representing Lesbian, Gay, Bisexual, Trans-gendered, Queer, Intersex, and Asexual peoples.

References

Albonetti, C. (1987), 'Prosecutorial discretion: The effects of uncertainty', *Law and Society Review*, **21** (2), 291–313.

Alexander, Michelle (2012), *The New Jim Crow: Mass Incarceration in the Age of Colorblindness*, New York: The New Press.

Barkow, R. (2010), 'Organizational guidelines for the Prosecutor's Office', *Cardozo Law Review*, **31** (6), 2089–2118.

Bazelon, Emily (2019), *Charged: The New Movement to Transform American Prosecution and End Mass Incarceration*, New York: Random House.

Beckett, Katherine (1997), *Making Crime Pay: Law and Order in Contemporary American Politics*, Oxford, UK: Oxford University Press.

Blomley, N. and J. Bakan (1992), 'Spacing out: Towards a critical geography of law', *Osgoode Hall Law Journal*, **30** (3), 661–690.

Bordenkircher v. *Hayes*, 434 U.S. 357 (1978).

Braverman, Irus, Nicholas Blomley, David Delaney and Alexandre Kedar (eds.) (2014), *The Expanding Spaces of Law: A Timely Legal Geography*, Palo Alto, CA: Stanford University Press.

Buzan, Barry (1991), *People, States and Fear: An Agenda for International Studies in the Post-Cold War Era*, 2nd edition, Boulder, CO: Lynne Rienner.

Chataway, M., T. Hart, R. Coomber and C. Bond (2017), 'The geography of crime fear: A pilot study exploring event-based perceptions of risk using mobile technology', *Applied Geography*, **86**, 300–307.

Color of Change (2019), 'Manhattan DA Cy Vance: Reopen the Linda Fairstein Cases', accessed 10 October 2019 at https://act.colorofchange.org/sign/reopen_linda _fairstein_cases?source=coc_main_website

Davis, A. (1998), 'Prosecution and race: The power and privilege of discretion', *Fordham Law Review*, **67** (1), 13–68.

Delaney, D. (2016), 'Legal geography II: Discerning injustice', *Progress in Human Geography*, **40** (2), 267–274.

England, M. and S. Simon (2010), 'Scary cities: Urban geographies of fear, difference and belonging', *Social & Cultural Geography*, **11** (3), 201–207.

Ferraro, Kenneth (1995), *Fear of Crime: Interpreting Victimization Risk*, Albany: SUNY Press.

Fischer, A. G. (2019), 'Linda Fairstein is under fire for the Central Park Five. But another part of her career deserves greater scrutiny', *The Washington Post*, 12 June, accessed 10 October 2019 at https://www.washingtonpost.com/outlook/2019/06/12/linda-fairstein-is-under-fire-central-park-five-another-part-her-career-deserves-greater-scrutiny/

Gershman, B. (1993), 'A moral standard for the prosecutor's exercise of the charging discretion', *Fordham Urban Law Journal*, **20**, 513–530.

Ghandnoosh, N. (2014), 'Race and punishment: Racial perceptions of crime and support for punitive policies', report released by The Sentencing Project, Washington, DC, USA.

Glaze, L. and E. Haberman (2013), 'Corrections: Key facts at a glance', report released by the Bureau of Justice Statistics, Washington, DC, USA.

Godsey, M. and T. Pulley (2003), 'The innocence revolution and our evolving standards of decency in death penalty jurisprudence', *University of Dayton Law Review*, **29**, 265–292.

Grawert, A. and C. Kimble (2019), 'Crime in 2018: Final analysis', report released by The Brennan Center for Justice, New York City, NY, USA.

Green, B. (2019), 'Prosecutorial discretion: The difficulty and necessity of public inquiry', *Dickinson Law Review*, **123** (3), 589–626.

Green, B. and E. Yaroshefsky (2016), 'Prosecutorial accountability 2.0', *Notre Dame Law Review*, **92**, 51–116.

Gregory, Derek and Allan Pred (2007), *Violent Geographies: Fear, Terror, and Political Violence*, London, UK: Taylor & Francis Press.

Henson, B. and B. Reyns (2015), 'The only thing we have to fear is fear itself ... and crime: The current state of the fear of crime literature and where it should go next', *Sociology Compass*, **9** (2), 91–103.

Hetey, R. C. and J. L. Eberhardt (2018), 'The numbers don't speak for themselves: Racial disparities and the persistence of inequality in the criminal justice system', *Current Directions in Psychological Science*, **27** (3), 183–187.

Hinton, Elizabeth (2018), *From the War on Poverty to the War on Crime: The Making of Mass Incarceration in America*, Cambridge, MA: Harvard University Press.

Jackson, R. H. (1940), 'The federal prosecutor', *Journal of the American Judicature Society*, **24**, 18–20.

Kahan, D. (2011), 'Neutral principles, motivated cognition, and some problems for constitutional law', *Harvard Law Review*, **125**, 1–77.

Kohler-Hausmann, Issa (2018), *Misdemeanorland: Criminal Courts and Social Control in an Age of Broken Windows Policing*, Princeton, NJ: Princeton University Press.

Law, V. (2014), 'Against carceral feminism', *Jacobin Magazine*.

MacLean, C., J. Berles and A. Lamparello (2015), 'Stop blaming the prosecutors: The real causes of wrongful convictions and rightful exonerations', *Hofstra Law Review*, **44**, 151–200.

Marshall, L. (2003), 'The innocence revolution and the death penalty', *Ohio State Journal of Criminal Law*, **1** (2), 573–584.

McDonald, Lynn (2019), *Sociology of Law & Order*, London, UK: Routledge.

Medwed, D. (2010), 'Emotionally charged: The prosecutorial charging decision and the innocence revolution', *Cardozo Law Review*, **31**, 2187–2214.

Miller, Lisa Lynn (2016), *The Myth of Mob Rule: Violent Crime and Democratic Politics*, Oxford, UK, Oxford University Press.

Miller, M. (2004), 'Domination and dissatisfaction: Prosecutors as sentencers', *Stanford Law Review*, **56**, 1211–1252.

New York City Press Office (2014), 'Statements of Mayor de Blasio and Corporation Counsel Zachary W. Carter on Central Park Five Settlement', 5 September, accessed 10 October 2019 at https://www1.nyc.gov/office-of-the-mayor/news/431-14/statements-mayor-de-blasio-corporation-counsel-zachary-w-carter-central-park-five

Nixon, Rob (2011), *Slow Violence and the Environmentalism of the Poor*, Cambridge, MA: Harvard University Press.

O'Lear, Shannon (2018), *Environmental Geopolitics*, Lanham, MD: Rowman & Littlefield.

Pettit, B. and C. Gutierrez (2018), 'Mass incarceration and racial inequality', *The American Journal of Economics and Sociology*, 77 (3–4), 1153–1182.

Pfaff, John (2017), *Locked In: The True Causes of Mass Incarceration and How to Achieve Real Reform*, New York: Basic Books.

Reiz, N., S. O'Lear and D. Tuininga (2018), 'Exploring a critical legal cartography: Law, practice, and complexities', *Geography Compass*, 12 (5), 1–10.

Roach, Kent (2019), 'Regulating the prosecutorial role in wrongful convictions', in Victoria Colvin and Philip Stenning (eds.), *The Evolving Role of the Public Prosecutor: Challenges and Innovations*, New York: Routledge, pp. 249–281.

Schuermans, N., B. Meeus and F. De Maesschalck (2010), 'Is there a world beyond the Web of Science? Publication practices outside the heartland of academic geography', *Area*, 42 (4), 417–424.

Simon, Jonathan (2007), *Governing Through Crime: How the War on Crime Transformed American Democracy and Created a Culture of Fear*, Oxford, UK: Oxford University Press.

Sklansky, D. (2016), 'The nature and function of prosecutorial power', *Journal of Criminal Law & Criminology*, 106, 473–520.

Smith, R. and J. Levinson (2011), 'The impact of implicit racial bias on the exercise of prosecutorial discretion', *Seattle University Law Review*, 35, 795–826.

Spade, Dean (2015), *Normal Life: Administrative Violence, Critical Trans Politics, and the Limits of Law*, Durham, NC: Duke University Press.

Spohn, C. (2018), 'Reflections on the exercise of prosecutorial discretion fifty years after publication of *The Challenge of Crime in a Free Society*', *Criminology & Public Policy*, 17, 321–340.

Spohn, C., K. Tellis and E. N. O'Neal (2014), 'Policing and prosecuting sexual assault', *Critical Issues on Violence Against Women: International Perspectives and Promising Strategies*, 3, 93–103.

Springer, S. and P. Le Billon (2016), 'Violence and space: An introduction to the geographies of violence', *Political Geography*, 52, 1–3.

Stolzenberg, L., S. D'Alessio and D. Eitle (2013), 'Race and cumulative discrimination in the prosecution of criminal defendants', *Race and Justice*, 3 (4), 275–299.

Taylor, R., R. Miller, D. Mouzon, V. Keith and L. Chatters (2018), 'Everyday discrimination among African American men: The impact of criminal justice contact', *Race and Justice*, 8 (2), 154–177.

The National Registry of Exonerations (2019), accessed 21 September 2019 at https://www.law.umich.edu/special/exoneration/Pages/about.aspx

Tonry, M. (2012), 'Prosecutors and politics in comparative perspective', *Crime and Justice*, 41 (1), 1–33.

Walker, Samuel (2006), *Sense and Nonsense about Crime and Drugs*, Belmont, CA: Thomson Wadsworth Press.

Ward, G. (2014), 'The slow violence of state organized race crime', *Theoretical Criminology*, **19** (3), 299–314.

Weaver, V. (2007), 'Frontlash: Race and the development of punitive crime policy', *Studies in American Political Development*, **21** (2), 230–265.

Webster, E. (2019), 'A Postconviction Mentality: Prosecutorial Assistance in Exoneration Cases', *Justice Quarterly*, **36** (2), 323–349.

Wechsler, H. (1959), 'Toward neutral principles of constitutional law', *Harvard Law Review*, **73** (1), 1–35.

When They See Us (2019), dir. Ava DuVernay, Netflix.

Wright, R. and M. Miller (2010), 'The worldwide accountability deficit for prosecutors', *Washington & Lee Law Review*, **67**, 1587–1622.

Wyn Jones, Richard (1999), *Security, Strategy, and Critical Theory*, Boulder, CO: Lynne Rienner Press.

Zalman, M. and R. Grunewald (2015), 'Reinventing the Trial: The Innocence Revolution and Proposals to Modify the American Criminal Trial', *Texas A&M Law Review*, **3**, 189–260.

Zhao, J., B. Lawton and D. Longmire (2015), 'An examination of the micro-level crime–fear of crime link', *Crime & Delinquency*, **61** (1), 19–44.

13 Dark cartographies: mapping slow violence

Peter Vujakovic

Introduction: maps and cultural violence

Slow violence is sustained by cultural violence – 'the symbolic sphere of our existence' – used to normalise oppression (Galtung, 1990, p. 291). Maps have been used extensively to control people and resources and manipulate worldviews. Maps as tools of spatial knowledge and control play a major part in sustaining violence, from waging war and 'policing actions', to the slow violence inherent in most socio-economic systems. While maps are generally the products of elites, counter-mappings can challenge slow violence.

A personal example exemplifies the power of maps. In 2014 an entry entitled 'The power of maps' was published in *The Times Comprehensive Atlas of the World* (Times Books, 2014, pp. 42–43). The article discussed Harley's (2001) concept of political silences in maps, whereby 'the poor and marginalised in society have been ignored' (p. 43). The atlas article noted that silences are common in mapping, for instance, the 'removal' of inconvenient settlements or strategic sites, for example the invisibility of African townships such as Soweto on official maps from the Apartheid era (p. 43) in South Africa. This prompted a reader's rebuke:

> This is patently wrong: within a few minutes I pulled out a 1982 1:2 5000 000 [sic] scale Surveyor General map of South Africa which clearly shows Soweto and Tembisa. If I had the inclination and the time I could pull out several Surveyor General maps from the 70s and 80s which show townships where the scale is appropriate. How an error like this, which could have been avoided with minimal research, reached the publication stage of the 14th edition is difficult to understand. (From communication to the publisher)

As the author of the entry being criticised, I was asked how we might reply. I responded that my information came from Stickler (1990) who provided examples of small-scale maps used in education, travel and tourism in South Africa that did not include Soweto, despite being roughly twice the population of nearby Johannesburg. Invisibility was greater still for informal settlements, 'Many maps of South Africa are characterised by the deliberate elimination not of small towns but of Black towns' (p. 329). 'Black townships' were omitted or downgraded; for example, 'White towns' of the eastern Witwatersrand were consistently shown more prominently than adjacent 'Black towns' despite being approximately four times smaller in population terms. Cartography was clearly a tool of cultural violence.

Nevertheless, the reader was right in pointing out that detailed state topographic mapping did include 'Black' settlements. However, this is no surprise, as it would be necessary to have detailed mapping for effective 'police actions' and other forms of socio-economic control. Modern state topographic mapping emerged from war and has always been used in diverse ways, some benign, some brutal.

This anecdote is a reminder of two fundamental types of map, each contributing to cultural and slow violence: topographic maps and thematic maps. Topographic maps are general use maps, providing an overview of the physical and human geography of a region, including administrative units as well as physical features. Thematic maps, often used in education, focus on one or more key issues (e.g. population distribution, infant mortality rates) and generally lack the detail of topographic maps. These two forms are not mutually exclusive, hybrids are common, for example detailed geology and land-use maps superimposed on topographic maps. Both exist at a range of scales, but thematic maps are generally small-scale while topographic map are larger scale, providing information for multiple purposes from deployment of armed forces to infrastructure development and recreational uses. Both forms are discussed below.

This chapter explores the map as a tool of control, both directly as sources of spatial information and in their use in propagating worldviews that sustain the status quo. Counter-mapping is considered as a means of combating slow violence.

Maps as a 'smoking gun': cartography and slow violence

... many of us are not aware and most of us not sufficiently aware that *maps are weapons*. (Weigert, 1941, p. 528, emphasis in the original)

Many tools and technologies can be put to both benign and malign uses. Maps are no different. Weigert (1941) understood the role of the map as 'a psychological weapon in a warring world' (p. 528), specifically its use in Nazi Germany to normalise the expansionist vision of *Lebensraum*.

Understanding the political and social significance of maps and the cartographic process has undergone a series of paradigm shifts. From a focus on communication models via post-modernist/deconstructionist approaches, to recent discussion of a post-representational cartography. Much recent debate has treated maps as 'texts', providing valuable insight into the power of maps. Viewing the map as a 'tool', crafted for a purpose, is also helpful.

The cartographic communication paradigm attempted to understand how maps transmit objective spatial information and the legacy of this approach remains (Board, 2018); not surprising given that many cartographers emerge from a 'scientific' training in geographic information systems (GIS). This, however, masks the socio-political significance of mapping among its practitioners who are embedded in institutional settings, and creates an unwarranted 'trust' in maps in society.

While the propaganda value of maps is long recognised (Quam, 1943; Soffner, 1942; Thomas, 1949; Weigert, 1941), the idea that *all maps* serve an interest, even if that interest is veiled, is more recent and founded on the work of Brian Harley (see *The New Nature of Maps* (2001) a posthumous collection of Harley's key papers) and Wood (1992). Harley draws attention to the 'hidden power of cartography' (embedded within its apparent objectivity) that takes two forms: 'external power' exerted on and through cartography, and 'internal power', the map as representation of order and status. These critical examinations of cartography build on post-modern deconstructionist approaches to 'texts'. As Harley noted '[d]econstruction urges us to read between the lines of the map ... to discover the silences and contradictions that challenge the apparent honesty of the image' (Harley, 2001, p. 153). Harley was largely concerned with historic maps, for example European settlement maps of America and their use to suppress Indigenous geographies. Others have focused on contemporary themes in use of cartography in hegemonic discourses; for example, contested geographic identities in the Middle East (Culcasi, 2011, 2012; Leuenberger and Schnell, 2010). Kitchen and Dodge (2007) go further and argue for a post-representational view of cartography in which every time a map is engaged with it is newly interpreted and made to do work, it is in a constant state of becoming (see Azócar et al., 2014, for a detailed review of paradigm shifts in cartographic research).

This chapter argues that to treat maps and mapping as either objective representations of the world or at the other extreme as inherently unstable endangers critical examination of the work that maps do and distract from understanding how they can act as viral memes that constantly underpin dominant tropes. For example, every time a Eurocentric world map is engaged with it reinforces a western hegemonic bias (Vujakovic, 2013); Galtung would certainly recognise this as a form of cultural violence.

Arguments about 'progress' in cartography are also part of current debate. Edney (1993), for example, contests the privileged position of western 'scientific' cartography, seeing value in other mappings. This, however, distracts from the fact that improvement in geo-spatial science constantly enhances the power of elites. Mary Kaldor's (1981) concept of the 'baroque arsenal' provides a nuanced understanding of the role of evolving technologies that can be applied to cartography and geo-spatial sciences more generally as a 'tool-kit' of control. As Kaldor notes:

> Baroque technological change represents "improvements" to successive weapons systems which can pass through phases of invention, innovation, and integration *without disturbing the social organization of the users.* (Kaldor, 1986, p. 591, emphasis added)

Baroque modalities seek to maintain the existing order, while also modernising (Del Valle, 2002). The notion of 'baroque' as a system of regulation was elucidated by Maravall in his classic interpretation of the Spanish Baroque as a 'structure' of social control (Maravall, 1975, trans. 1986). Much of the social control of the baroque was mediated by spectacle, a concept revisited by Debord (1967, trans. 2009) in his 'society of spectacle'; this comprises a culture of compliant consumption in which spectacle 'monopolizes the majority of the time spent outside the production process' (p. 25), and deference to state's processes deemed important in matters related to threat and terror (e.g. increasing surveillance and intrusion).

Cartography may appear benign when compared to the military systems discussed by Kaldor but it can be seen as part of a baroque arsenal; it must be remembered that geo-spatial science underpins both the overt violence of warfare and the complex infrastructures supporting structural, slow and cultural violence. To paraphrase Kaldor, a baroque cartography takes advantage of phases of invention and technical innovation without disturbing the socio-economic organization of the elites that control it (Vujakovic 2018a).

Maps, power and the state

> As much as guns and warships, maps have been weapons of imperialism ... a tool of pacification, civilization, and exploitation. (Harley, 2001, p. 57)

Historically map production and consumption have been the prerogative of the powerful; they have been used to awe, to shape worldviews, and to control peoples and resources. The map as a tool of power comes fully into its own during the Enlightenment era along with the evolution of the modern state and with advances in technologies of survey and reproduction.

Attention has been drawn to the critical connection between the power of mapping and the notion of territory, especially in the rise of the nation-state (Wood, 2005). For some, the map *is* the territory, in the sense that the act of mapping brings a territory into being: '... maps conjured up borders were none had existed' (p. 33). The modern 'nation-state' originated in Europe in the nineteenth century based on the concept of a distinctive 'people' occupying and controlling a *defined territory*. Mapping defines the state as a discreet entity, some by default. Siam, for example, emerged as a 'territorial-state' by being enclosed by neighbouring colonial 'borders' (Anderson, 2006). European mapping came to shape the geopolitical imagination globally.

Conceiving the territorial state as a 'power container' (a 'sovereign power in a particular territory', Taylor, 1994, p. 151) aids understanding of how maps contribute to cultural violence, and how elites use maps to control resources and construct hegemonic worldviews. International boundary lines are of major significance to the state; they not only represent the limits of jurisdiction and ownership of resources, but also have immense psychological significance. State elites use maps to control the contents of the container and, in some cases, to project power globally. Once the importance of the relationship between the map, the territorial-state and the power-container is understood, it becomes clear why maps are a powerful form of representation.

As noted, mapping is divided between topographic and thematic formats. Topographic maps, especially detailed large-scale maps, involve a massive investment but are essentially to the management of the state, from benign purposes such as delivering utilities, to forms of slow violence such as 'red-lining' (the systematic denial of services by government agencies and businesses to residents of defined neighbourhoods).

Modern state mapping emerged in various ways. The principles that underpinned state mapping and still inform online geo-spatial applications today, evolved from 'the methods of triangulation and geodetic measurements first proposed and practiced by the Cassinis' (Brotton, 2013, p. 296) to map France. The Cassinis were a family of scientists who contributed to the first modern mapping of a nation. Cassini's mapping project and its entanglement with the end of the French monarchy and the first Republic is a story of politics as much as science. Brotton notes that despite 'initial attempts to establish a way of mapping that could police and control the dynastic realm, the monarchy supported the map of a kingdom which unintentionally metamorphised into the map of a nation' (p. 336). A sense of national identity emerged when people could see an image of the nation.

The British Ordnance Survey (OS; initiated in 1791 (Hodson, 1993)) and other state mapping enterprises have facilitated the ability of elites to impose their will on the land, its people and resources. Perhaps the most famous of state mapping initiatives, the OS was born of the Military Survey of Scotland that followed the violent suppression of the Jacobite Rebellion of 1746. The original survey enabled the slow violence of fortification, clearance and enclosure of the Scottish Highlands, and reorganisation of land confiscated from Jacobite landowners. As Kaplan (2018) notes:

> The First Military Survey of Scotland sought to represent vanquished terrain in accurate detail. Moving through a devastated landscape of retributive state violence, the British Board of Ordnance surveyors produced a vertical view of military occupation, representing a formidable terrain as largely "blank". (p. 31)

Earlier still than the OS, the British 'Survey of India' (initiated 1767) helped consolidate the hold of the British East India Company on its territories and later British colonial supremacy. But the maps prepared during the colonial period, that inevitably portrayed a British image of India, also paved the way for Indian nationhood (Mondal, 2019) in a similar manner to the Cassini map of France. Maps also served the extra-territorial projection of power as well as aiding the colonial enterprise, internal or overseas. The most significant mapping project in history was the Soviet Union's world-mapping initiative, conducted by its Military Topographic Directorate; this produced over a million detailed maps of different parts of the world. These have only been circulated widely since the downfall of the Soviet Union. It has been suggested that these maps were not primarily designed to facilitate military activity, but rather to support administration, perhaps following a coup (Davies and Kent, 2017).

Maps and innovations in geo-spatial information systems continue to facilitate the projection of power. Unmanned Combat Air Vehicles (UCAVs) – 'military drones' – are perhaps the ultimate expression of the baroque arsenal, facilitated by baroque cartographies. These are regarded as ideal weapons for the remote, extra-territorial delivery of death, saving time and operator lives and improving operational ability (Coker, 2009). As well as their destructive force, they also generate novel forms of violence. As Barrinha and da Mota (2017) note 'drones produce a whole new security environment in which people are supposed to "normally" go about their lives, while being in the same geographic areas in which the US conducts its … strikes' (pp. 7–8). They cite Chamayou (2015, p. 45) on the psycho-spatial impact:

> [drones] inflict mass terror upon entire populations. It is this […] that is the effect of permanent lethal surveillance: it amounts to a psychic imprisonment within a perimeter no longer defined by bars, barriers, and walls, but by the endless circling of flying watchtowers up above. (Cited on p. 12)

And even the servants of these systems are not immune to the violence engendered. A study of UCAV operators for the US military showed no significant difference in the rates of mental health diagnoses compared with pilots overflying combat zones (Otto and Webber, 2013, p. 3) Exposure to combat events, particularly those involving death of civilians, causes significant trauma (Chappelle et al., 2019).

As well as national scale mapping, local mapping can serve as a tool of oppression or discrimination. Following the Great Depression, for example, the US federal government created the Home Owners' Loan Corporation (HOLC) to control the housing market. The HOLC created loan security maps for residential areas of many cities, graded in terms of lending risk. It is argued that this may have contributed to long-standing variations in neighbourhood development. The lowest-rated areas were drawn in red, often areas with large populations of African Americans. This led to the practice of 'red-lining' by which people were denied access to credit (Aaronson et al., 2017). Aaronson and his co-authors found that neighbourhoods 'red-lined' 'in the 1930s experienced a marked increase in racial segregation in subsequent decades that peaked around 1970 before beginning to decline' (p. 1).

Maps were also important in so-called 'urban triage' policies in New York City in the 1970s; where many neighbourhoods had been in decline from the late 1960s. NYC's housing administrator, Roger Starr, advocated 'planned shrinkage', withdrawal of services (e.g. closure of firehouses, schools and subway stations) in low-income and ethnic minority neighbourhoods that Starr believed could not be saved (Aalbers, 2014a). This planned shrinkage impacted large areas of the city, including parts of the Bronx, Brooklyn, Lower East Side and Harlem. Aalbers provides examples of maps designed to facilitate and rationalise closure of fire services from a document by Kolesar and Walker (1972). Aalbers (2014b) suggests that influential maps produced as part of plans for the rebuilding of New Orleans following hurricanes Katrina and Rita in 2005 appeared 'like urban triage all-over again: neighbourhoods that have been heavily damaged should be depopulated and will not be rebuilt' (p. 563) in a city where much of the population lives below the poverty line.

The cultural violence of toponyms

Cultural violence is also reinforced by place-naming (or removal) and inscription of these toponyms on maps. Place-naming claims space; repetition, standardization and displacement of former names, normalises it. Names endow place with new meaning in settler and colonial situations – an assertive 'violent geography' (Peteet, 2005). As Harley (2001) notes:

> Place names have always been implicated in the cultural identity of people who occupy the land. Naming a place anew is a widely documented act of political possession in settlement history. Equally, the taking away of a name is an act of dispossession. (pp. 178–179)

Naming reinforces the 'sense of place' that privileges the European and denigrates Indigenous peoples. Harley focused on New England during the seventeenth century, but other examples exist and persist today. In the same century Irish place-names, regarded as 'barbarous' by Charles I, were to be expunged (later restored by the OS, in part as an antiquarian exercise (Close, 1969)). Barber (1993a) describes how Germany overprinted German place-names on maps of Czechoslovakia, even in districts without German-speaking populations: 'They bloodlessly implied expulsion, misery, cultural and probably physical genocide' (p. 83).

The Arab-Israeli conflict provides contemporary examples. Cohen and Kliot (1992) discuss the symbolic role of place-name mapping in disputed territory;

the selection of Israeli place-names became a tool for reinforcing Zionist ideologies, especially of land captured in the Six-Day War (1967). Naming generates symbolic meaning, by evoking modern 'heroes' or the ancient Israelite kingdom, it creates a Jewish landscape (e.g. reference to the occupied West Bank by the biblical names of 'Judea' and 'Samaria' (Peteet, 2005)). The Palestine Liberation Organization (PLO) countered by copying Israeli maps, erasing the Hebrew typography and overprinting with Arabic text and place-names (Kadmon, 2004).

Once a place is named it is difficult to change, often for practical reasons, inscription on maps, gazetteers, signage, letterheads, etc. (Monmonier, 2006), although it is clearly possible with sufficient political will. Post-colonial landscapes are an obvious example. Njoh (2017) discusses how place-naming was 'a tool for articulating power in Anglophone and Francophone Africa' (p. 1174). With the passing of colonialism independent states reacted differently: Kenya quickly replaced Eurocentric with Afrocentric place-names, while Senegal remained wedded to place-names from the 'Eurocentric cultural lexicon'. In the late 1980s the US Board of Geographic Names renamed numerous sites containing elements deemed racist (Loewen, 1999), but many derogatory names remain. Online searches show that these names are still extant. Examples included Squaw Tits and Squaw Tit about 450 kilometres east of San Diego which can be seen on Google Earth and OpenStreetMap (search date 5 June 2020). See Rose-Redwood et al. (2010) for a review of critical place-name studies.

Naming of places, like statues and other symbolic representations of power, contributes to sustained cultural violence and feelings of subjugation. The recent Black Lives Matter movement is a timely expression of a revulsion for the continuance of such symbols.

Maps as spectacle

We live in a map-immersed world of 'spectacle' in which neo-baroque cartographies manufacture infographics that mesmerise, distract and sometimes paralyse engagement with social and environmental issues (Vujakovic, 2018a). Aided by developments in design technologies these infographics have morphed into 'infotainment' in which spectacle overrides serious content. The concept of the baroque as a near-seamless web of control, together with the concept of spectacle, is even more pertinent in a world of 'digital-zombies' trapped in the bubble of their personal electronic devices: 'The Spectacle is the growing Nothing in the lifeblood of society' (Smith, 2014, p. 14).

As well as creating compliant consumers, Debord (1998) suggests spectacle can be mobilised to create compliance through fear; the spectator must never know too much but always enough to convince them that any means authority evokes to control social, geopolitical and environmental terrors is acceptable. Maps in the news media are perhaps the most important form of thematic mapping that most people are exposed to in this respect. In the maps accompanying reportage of the London terrorist bombings of 2005, for example, *The Times* (8 July) turned the iconic tube map and a mundane street map of the capital into a landscape of fear with a combination of explosion icons and inset diagrams showing detonations on buses and trains.

If, as Monmonier (1989) has stated, the news media are society's most influential geographic educator and cartographic gatekeeper, then the type and quality of news maps produced must have some direct bearing on geographic understanding. News maps act at a fundamental level in terms of influencing public opinion by creating specific connotations relating to places and events, for example, like the 'red-lining' discussed above, news maps can be part of the process by which places are labelled in negative ways, linked to economic decline, crime or environmental quality.

The popular (mis)understanding of maps as objective images can also lead to their use as visual metaphors for 'accuracy' and 'precision', connotations that supported NATO's scripting of 'surgical strikes' as a legitimate and humane action against Yugoslavia during the 1999 Kosovo crisis, acting 'as both guarantor and visual confirmation of the ability of "our" forces to hit the right targets' (Vujakovic, 2002, p. 192). When refugees were accidentally killed the metaphor failed and the persistent use of maps showing bombing of military related targets (55% of *all maps* in the UK quality press in April 1999) suddenly evaporated. This example is a specific form of Harley's (2001) 'hidden power' of cartography, 'internal power' as a visual representation of order and status. Unfortunately, as print and screen technologies have developed, the baroque nature of information graphics as a spectacle has become more obvious. Information has morphed into 'infotainment'. Clarity is lost in what Tufte (2006) has called 'chartjunk'. Classic examples include use of maps in threat discourse involving North Korea and China missile systems. These have been some of the most spectacular maps in the news media in recent years, with pyrotechnic displays as part of drama; for example, *The Times* map of North Korea's missile threat (13 October 1999, p. 11); in this graphic a US 'hit-to-kill' missile blasts a Korean delivery system from the sky in a fireball of flame, over a bizarre map of the world with a false horizon and missile ranges that make no sense at all – drama supersedes accuracy in this example of 'infotainment' (Vujakovic, 2014).

'Oughtness mapping' – maps of how things are and how things ought to be

Commissioning maps is generally the preserve of the powerful, designed to serve their interests, sometimes benign, but often not. Subversion of the map's purpose does, however, occur. An early example relates to a map of Barbados created in the late seventeenth century; its maker, Richard Forde, was, according to Barber (1993b), 'a rare exception to the general rule that map-makers reflect the values of the establishment in their work' (p. 31). The first economic map of an English American colony it represented key sites in detail, even to the extent of differentiating between wind, water and cattle-powered mills used to grind sugar cane produced by a slave economy. Despite giving some support to this oppressive system, Forde, a Quaker, did manage to subvert the map by leaving the colonial churches and numerous coastal fortifications unmapped, much to the anger of the Governor and against the instructions of the crown.

Forde's map could be regarded as an early, if veiled, form of critical cartography. Perkins (2018) distinguishes between two forms of 'critical cartography': first, 'theory' – new ways of writing about maps, and, second, 'practice' – cartographies of social concern that seek to reveal the spatial inequalities and injustices hidden by conventional mapping practices. The latter can be further sub-divided into maps that reveal slow violence and those that provide alternative views of the world – 'counter-hegemonic cartography' (Kitchen et al., 2011).

Perkins traces the practice of critical cartography to William Bunge's radical mappings of US cities in the 1960s. Bunge's concern for social injustice led to the 'Detroit Geographic Expedition' that aimed to bring academics together with 'folk geographers' (people with no formal training) and African Americans to generate '"oughtness maps" – maps of how things are and maps of how things ought to be' (kanarinka, 2013; unpaginated). Bunge's mapping captured both elements – revealing slow violence and offering alternatives – of practical critical cartography.

Bunge's classic, *Nuclear War Atlas* (1988) developed many themes derived from his 'expeditions' regarding the lived experience of disadvantaged urbanites, particularly children. It contains some simple but powerful maps, including his stark black and white map of infant mortality rates across Detroit (p. 165); the impact comes not from the numbers but the fact that it displays these in terms of other countries; a very short drive of about 4 miles along Grand River Avenue revealed mortality rates that rose drastically from that of Norway (10.7 per thousand) to that of Guyana (43.9 per thousand). Bunge uses colour in other maps, but sparingly, red for 'death' and green for 'life'. These are maps that stick in the mind – 'Map 3.16 Region of rat-bitten babies' based on Detroit Department of Health data, along with Bunge's narrative: 'The rats – predacious carnivores – are looking for food to eat, be it garbage or infant's toes, fingers and nose' (p. 167). Bunge's maps are resonant of his radical 'grassroots neighbourhood insurrection' (Barnes, 2018, p. 1697). His approach had, however, its antecedent in the innovative work of others. Examples include Charles Booth's poverty maps of London and Otto Neurath's socially concerned 'picture education'.

Booth's 17-volume 'Life and Labour of the People of London' was published between 1889 and 1903 (Shepherd, 1993). His maps are early examples of sociological cartography, with streets coloured to indicate the social class and income of those living there; although the labelling is illustrative of contemporary opinions regarding the poor (Black = 'Lowest class. Vicious, semi-criminal') (LSE, 2016). It was a first step in 'oughtness mapping'. Twelve sheets were eventually published (1902–1903) covering much of the metropolis. Booth had originally set out to contest the figures for the urban poor presented by the Social Democratic Federation, only to find that it was far worse.

While Booth's maps revealed the results of slow violence, it is worth noting that his data were overprinted on Stanford's Library Map of London (1:10560), on which the grey areas representing non-residential land uses provide an insight into conditions in the capital. These, to a large extent, represented a landscape of power, control and surveillance, including penitentiaries, asylums and barracks (Wellington Barracks continues to provide protection for Buckingham Palace) and provided a subliminal reassurance for the 'well-off' with access to Booth's publication (an embedded 'cultural violence'?).

Neurath's educational graphics, by contrast, arose from his interest in public instruction in a semi-literate society and they were developed when he was responsible for organising popular expositions concerning social welfare in Vienna in the 1920s (Hartmann, 1997). Through his pictorial (ISOTYPE) approach, Neurath believed he had achieved a form of representation that 'had something of the unemotive quality of numbers but was fascinating enough to interest the layman' (Twyman, 1975, p. 13). Neurath's pictorial education has continued to influence a range of representations of poverty and human tragedy, for example a 2007 *New York Times* graphic displaying military and civilian casualty rates in Iraq (reprinted in Jansen, 2009). His legacy is clear in the pictorial methods used in a range of critical cartographies, for example the atlases published by Pan Books (in association with Pluto Press) from the 1980s, starting with Kidron and Segal's (1981) *The State of the World Atlas*. Their authors frequently used Neurath-like pictograms to display quantitative data. The first edition of the *Women in the World* atlas (Seager and Olson, 1986) clearly expressed their belief that graphics can provide an insight into issues that would otherwise remain 'obscure':

> By mapping the world of women, patterns are revealed that are usually obscured in statistical tables or in narratives. The similarities and differences, the continuities and contrasts among women around the world are best shown by – literally – mapping out their lives. (p. 7)

Others have followed, as has innovation in sociological mapping. An important example is Dorling and Thomas's (2011) *Bankrupt Britain: An Atlas of Social Change*, which illustrated the social recession that impacted much of the UK following the 2008 financial crisis using cartograms. These show locations proportional to their population rather than land area thereby drawing attention to issues in large cities which might otherwise be obscured (see Vujakovic, 1989, for discussion of equal-area versus cartograms for mapping development issues at the global scale). Unfortunately, the clarity of Booth's and Neurath's graphics has not always been maintained, and contemporary infographics are often marred by poor design. As Holmes (1996) notes in regard of ISOTYPE products '… the picture never got in the way of the information. Nor did the colour. Today's use of colour often hides the essential point of the graphic' (p. 49).

Oughtness mapping has also included radical cartographies designed to challenge the status quo; maps that display the viewpoint of marginalised groups and express their wishes (Parker, 2006; Perkins, 2007). It is not possible here to provide an in-depth discussion of community, participatory or collaborative mapping, but they have been used to reveal inequalities and challenge authority across a wide range of issues from urban and rural planning and environmental quality (Crouch and Matless, 1996; Anderson, 2011) and the interests of specific groups (e.g. disability access issues; Matthews and Vujakovic, 1995). Carton and Thissen's (2009) study of collaborative mapping in the Netherlands shows, however, that their use can deepen conflicts as competing interests are revealed. Some community mapping initiatives have also failed due to low levels of participation by residents (Parker 2006) and because they may serve the interests of conservative elements of a community (e.g. reinforce traditional conceptions of the rural idyll; Kent and Vujakovic, 2018).

In developing economies participatory mapping has become an important tool for achieving land rights and protecting natural resource. In Kenya, for example, two participatory mapping initiatives were undertaken in Kibera, one of largest informal settlements in Africa; the projects taught locals how to use GPS to produce topographic and thematic maps of the settlement (Panek and Sobotova, 2015). These were used to improve delivery of education, water, health care and sanitation. Mapping, while not without problems, helped the Kiberans to create a resource to contest dominant norms (Donovan, 2012); 'the residents of Kibera realised that "being on the map" in the digital world actually means "to exist"' (Panek and Sobotova, 2015, p. 4).

The Social Tenure Domain Model (STDM), developed by Global Land Tool Network (GLTN; an international collaboration supported by the UN), is a 'pro-poor information tool'. Communities identify spatial information on satellite images, air photographs, existing maps, or make sketch maps (Lemmen, 2013). Recent projects have included improving tenure security on customary lands in Zambia and work in Nepal to improve earthquake recovery, resilience and tenure security of communities affected by the April 2015 earthquake (GLTN, 2019). Such projects are, however, generally dependent on western aid, expertise, values and models which has caused concern (Kang'ethe and Manomano, 2014; see also the special issue of *The Cartographic Journal*, on participatory GIS, which explores ethical and practical issues regarding implementation and GIS's role in empowering of communities (Kar et al., 2016; editorial introduction)).

Conclusion

Despite developments in critical cartography, maps remain in many people's mind a symbol of objectivity and factual data, the product of scientific survey techniques which disguises their role in cultural and slow violence. Their role in control remains important, especially in a world in which the management of populations and resources becomes ever more critical with crises in the global economy, climate change and the impact of the coronavirus. Elites continue to build a baroque arsenal of tools from military to cultural to impose their will and extend control, and the map as spectacle dominates this format across the media.

Contemporary approaches towards a more human and humane counter-cartography are encouraging, but these alternative strategies will always be playing 'catch-up' with the continually evolving baroque technologies of control. To counter slow violence, it is critical that the tools of cultural violence are exposed and understood by as many people as possible.

References

Aalbers, Manuel B. (2014a), 'Do Maps Make Geography? Part 1: Redlining, Planned Shrinkage, and the Places of Decline', *ACME: An International E-Journal for Critical Geographies*, **13** (4), 525–556.

Aalbers, Manuel B. (2014b), 'Do Maps Make Geography? Part 2: Post-Katrina New Orleans, Post-Foreclosure Cleveland and Neoliberal Urbanism', *ACME: An International E-Journal for Critical Geographies*, **13** (4), 557–582.

Aaronson, Daniel, Daniel Hartley and Bhashkar Mazumder (2017), *The Effects of the 1930s HOLC "Redlining" Maps*, Working Paper, No. 2017-12, Chicago, IL: Federal Reserve Bank of Chicago.

Anderson, Benedict (2006, revised edition), *Imagined Communities: Reflections on the Origin and Spread of Nationalism*, London: Verso.

Anderson, Doug (2011), 'Community Mapping: Putting the Pieces Together', *The Geography Teacher*, **8** (1), 4–9.

Azócar, Fernandez, Iván Pablo and Manfred F. Buchroithner (2014), *Paradigms in Cartography: An Epistemological Review of the 20th and 21st Centuries*, Berlin: Springer-Verlag.

Barber, Peter (1993a), 'Cartographic Colonization', in P. Barber and C. Board (eds) *Tales from the Map Room: Fact and Fiction about Maps and their Makers*, London: BBC Books, pp. 82–83.

Barber, Peter (1993b), 'Sedition in the Colonies', in P. Barber and C. Board (eds) *Tales from the Map Room: Fact and Fiction about Maps and their Makers*, London: BBC Books, pp. 30–31.

Barnes, Trevor (2018), 'A Marginal Man and his Central Contributions: The Creative Spaces of William ("Wild Bill") Bunge and American Geography', *Environment and Planning A: Economy and Space*, **50** (8), 1697–1715.

Barrinha, André and Sarah da Mota (2017), 'Drones and the Uninsurable Security Subjects', *Third World Quarterly*, **38** (2), 253–269.

Board, Christopher (2018), 'The Communication Models in Cartography', in Alexander Kent and Peter Vujakovic (eds) *The Routledge Handbook of Mapping and Cartography*, Abingdon, UK: Routledge, pp. 29–43.

Brotton, Jerry (2013), *A History of the World in Twelve Maps*, London: Penguin Books.

Bunge, William (1988), *Nuclear War Atlas*, Oxford: Basil Blackwell.

Carton, L.J. and W.A.H. Thissen (2009), 'Emerging Conflict in Collaborative Mapping: Towards a Deeper Understanding?', *Journal of Environmental Management*, **90**, 1991–2001.

Chamayou, Grégoire (2015), *Drone Theory* (trans. Janet Lloyd), London: Penguin.

Chapin, Mac, Zachary Lamb and Bill Threlkeld (2005), 'Mapping Indigenous Lands', *Annual Review of Anthropology*, **34** (1), 619–638.

Chappelle Wayne, Tanya Goodman, Laura Reardon and Lillian Prince (2019), 'Combat and Operational Risk Factors for Post-Traumatic Stress Disorder Symptom Criteria among United States Air Force Remotely Piloted Aircraft "Drone" Warfighters', *Journal of Anxiety Disorders*, **62**, 86–93.

Close, Charles (1969) *The Early Years of the Ordnance Survey* (reprint of 1926 edition with new introduction by J.B. Harley), New York: Augustus M. Kelley Publ.

Cohen, Saul B. and Nurit Kliot (1992), 'Place-Names in Israel's Ideological Struggle over the Administered Territories', *Annals of the Association of American Geographers*, **82** (4), 653–680.

Coker, Christopher (2009), *War in an Age of Risk*, Cambridge: Polity Press.

Crouch, David and David Matless (1996), 'Refiguring Geography: Parish Maps of Common Ground', *Transactions of the Institute of British Geographers*, **2** (1), 236–255.

Culcasi, Karen (2011), 'Cartographies of Supranationalism: Creating and Silencing Territories in the "Arab Homeland"', *Political Geography*, **30**, 417–428.

Culcasi, Karen (2012), 'Mapping the Middle East from Within: (Counter-)Cartographies of an Imperialist Construction', *Antipode*, **44** (4), 1099–1117.

Davies, John and Alexander J. Kent (2017), *The Red Atlas: How the Soviet Union Secretly Mapped the World*, Chicago, IL: Chicago University Press.

Debord, Guy (1998), *Comments on the Society of the Spectacle*, London: Verso.

Debord, Guy ([1967] 2009), *Society of the Spectacle* (trans. K. Knabb), Eastbourne, UK: Soul Bay Press.

Del Valle, I. (2002), 'Jesuit Baroque', *Journal of Spanish Cultural Studies*, **3** (2), 141–163.

Donovan, K. (2012) 'Seeing Like a Slum: Towards Open, Deliberative Development', *Georgetown Journal of International Affairs*, **13** (1), 97–104.

Dorling, Daniel and Bethan Thomas (2011), *Bankrupt Britain: An Atlas of Social Change*, London: Policy Press.

Edney, Matthew H. (1993) '"Cartography without Progress": Reinterpreting the Nature and Historical Development of Mapmaking, *Cartographica*, **30** (2–3), 54–68.

Galtung, Johan (1990), 'Cultural Violence', *Journal of Peace Research*, **27** (3), 291–304.

GLTN (2019), *Social Tenure Domain Model Blog*, Global Land Tool Network, accessed at https://stdm.gltn.net/blog/

Harley, J. Brian (2001), *The New Nature of Maps: Essays in the History of Cartography* (posthumous collection edited by Paul Laxton), Baltimore, MD: The Johns Hopkins University Press.

Hartmann, Frank (1997), 'Speaking Signs', *Telepolis – Magazin der netzkultur*, accessed 10 September 2003 at https://www.heise.de/tp/features/Speaking-Signs-3412796.html

Hodson, Yolande (1993), 'Wartime Difficulties – Peacetime Mapping', in P. Barber and C. Board (eds) *Tales from the Map Room: Fact and Fiction about Maps and their Makers*, London: BBC Books, pp. 126–7.

Holmes, Nigel (1996), 'Are We Controlling Technology or is it Controlling Us?', in R. Houlkes (ed.) *Information Design and Infographics*, Rotterdam: European Institute for Research and Development of Graphic Communications, pp. 45–53.

Jansen, Wim (2009), 'Neurath, Arntz and ISOTYPE: The Legacy in Art Design and Statistics', *Journal of Design History*, **22** (3), 227–242.

Kadmon, Naftali (2004), 'Toponymy and Geopolitics: The Political Use – and Misuse – of Geographical Names', *The Cartographical Journal*, **41** (2), 85–87.

Kaldor, Mary (1981), *The Baroque Arsenal*, New York: Hill and Wang.

Kaldor, Mary (1986), 'The Weapons Succession Process', *World Politics*, **38** (4), 577–595.

kanarinka (aka Catherine D'Ignazio) (2013), 'The Detroit Geographic Expedition and Institute: A Case Study in Civic Mapping, MIT Center for Civic Media', accessed 3 March 2020 at https://civic.mit.edu/2013/08/07/the-detroit-geographic-expedition -and-institute-a-case-study-in-civic-mapping/

Kang'ethe, S.M. and Tatenda Manomano (2014), 'Exploring the Challenges Threatening the Survival of NGOs in Selected African Countries', *Mediterranean Journal of Social Sciences*, **5** (27), 1495–1500.

Kaplan, Caren (2018), *Aerial Aftermaths: Wartime from Above*, Durham, NC: Duke University Press.

Kar, Bandana, Renee Sieber, Muki Haklay and Rina Ghose (2016), 'Public Participation GIS and Participatory GIS in the Era of GeoWeb', *The Cartographic Journal*, **53** (4), 296–299.

Kent, Alexander and Peter Vujakovic (2018), 'Maps and Identity', in Alexander Kent and Peter Vujakovic (eds) *The Routledge Handbook of Mapping and Cartography*, Abingdon, UK: Routledge, pp. 413–426.

Kidron, Michael and Ronald Segal (1981), *The State of the World Atlas*, London: Pluto Press.

Kitchen, Rob and Martin Dodge (2007), 'Rethinking Maps', *Progress in Human Geography*, **31**, 331–344.

Kitchen, Rob, Martin Dodge and Chris Perkins (2011), *The Map Reader: Theories of Mapping Practice and Cartographic Representation*, Chichester, UK: Wiley-Blackwell.

Kolesar, Peter and Warren E. Walker (1972), *An Algorithm for the Dynamic Relocation of Fire Companies*, New York: The New York City RAND Institute.

Lemmen, Christiaan (2013), *The Social Tenure Domain Model: A Pro-Poor Land Tool*, International Federation of Surveyors, Copenhagen, accessed 3 June 2020 at https:// stdm.gltn.net/STDM_-_A_Pro_Poor_Land_Tool.pdf

Leuenberger, Christina and Izhak Schnell (2010), 'The Politics of Maps: Constructing National Territories in Israel, *Social Studies of Science*, **40** (6), 803–842.

Loewen, James W. (1999), *Lies Across America: What Our Historic Sites Get Wrong*, New York: The New Press.

LSE (2016), *Charles Booth's London Poverty Maps and Police Notebooks*, London School of Economics, accessed 27 October 2018 at https://booth.lse.ac.uk

Maravall, Jose A. ([1975] 1986), *Culture of the Baroque: Analysis of a Historical Structure* (trans. T. Cochran), Manchester, UK: Manchester University press (originally publ. 1975, as *La Cultura del Barroco*).

Matthews, M. Hugh and Peter Vujakovic (1995), 'Private Worlds and Public Places: Mapping the Environmental Values of Wheelchair Users', *Environment and Planning A*, **27**, 1069–1083.

Mondal, Tarun K. (2019), 'Mapping India since 1767: Transformation from Colonial to Postcolonial Image', *Miscellanea Geographica*, **23** (4), 210–214.

Monmonier, Mark (1989), *Maps with the News: The Development of American Journalistic Cartography*, Chicago, IL: University of Chicago Press.

Monmonier, Mark (2006), *From Squaw Tit to Whorehouse Meadow: How Maps Name, Claim and Inflame*, Chicago, IL: University of Chicago Press.

Njoh, Ambe J. (2017), 'Toponymic Inscription as an Instrument of Power in Africa: The Case of Colonial and Post-colonial Dakar and Nairobi', *Journal of Asian and African Studies*, **52** (8), 1174–1192.

Otto, Jean L. and Braynat J. Webber (2013), 'Mental Health Diagnoses and Counseling Among Pilots of Remotely Piloted Aircraft in the United States Air Force', *Medical Surveillance Monthly Report*, **20** (3), 3–8.

Panek, Jiri and Lenka Sobotova (2015), 'Community Mapping in Urban Informal Settlements: Examples from Nairobi, Kenya', *The Electronic Journal of Information Systems in Developing Countries*, **68** (1), 1–13.

Parker, Brenda (2006), 'Constructing Community through Maps? Praxis in Community Mapping', *Professional Geographer*, **58** (4), 470–484.

Perkins, Chris (2007), 'Community Mapping', *Cartographic Journal*, **44** (2), 127–137.

Perkins, Chris (2018), 'Critical cartography', in Alexander Kent and Peter Vujakovic (eds) *The Routledge Handbook of Mapping and Cartography*, Abingdon, UK: Routledge, pp. 80–89.

Peteet, Julie (2005), 'Words as Interventions: Naming in the Palestine–Israel Conflict', *Third World Quarterly*, **26** (1), 153–172.

Quam, Louis O. (1943), 'The Use of Maps in Propaganda', *Journal of Geography*, **42**, 21–32.

Rose-Redwood, Reuben, Derek Alderman and Maoz Azaryahu (2010), 'Geographies of Toponymic Inscription: New Directions in Critical Place-Name Studies', *Progress in Human Geography*, **34** (4), 453–470.

Seager, Joni and Ann Olson (1986), *Women in the World Atlas*, Bristol, UK: Pluto Press.

Shepherd, John (1993), 'Poverty and Comfort', in P. Barber and C. Board (eds) *Tales from the Map Room: Fact and Fiction about Maps and their Makers*, London: BBC Books, pp. 146–147.

Smith, Peter (2014), *On Walking*, Axminster, UK: Triarchy Press.

Soffner, Heinz (1942), 'War on the Visual Front', *American Scholar*, **11**, 465–476.

Stickler, Philip J. (1990), 'Invisible Towns: A Case Study in the Cartography of South Africa', *GeoJournal*, **22** (3), 329–333.

Taylor, Peter J. (1994), 'The State as Container: Territoriality in the Modern World-System', *Progress in Human Geography*, **18** (2), 151–162.

Thomas, Louis B. (1949), 'Map as Instruments of Propaganda', *Surveying and Mapping*, **9** (2), 75–81.

Times Books (2014), *The Times Comprehensive Atlas of the World* (14th edition), Glasgow, UK: HarperCollins.

Tufte, Edward R. (2006) *Beautiful Evidence*, Cheshire, CT: Graphics Press.

Twyman, Michael (1975), *The Significance of Isotype, in Graphic Communication through ISOTYPE*, Reading, UK: University of Reading.

Vujakovic, Peter (1989), Mapping for World Development', *Geography*, **74** (2), 97–105.

Vujakovic, Peter (2002), 'Mapping the War Zone: Cartography, Geopolitics and Security Discourse in the UK Press', *Journalism Studies*, **3** (2), 187–202.

Vujakovic, Peter (2013), 'Warning! Viral Memes can Seriously Alter your Worldview', *Maplines*, Spring 2013, pp. 4–6.

Vujakovic, Peter (2014), 'The State as a "Power Container": The Role of News Media Cartography in Contemporary Geopolitical Discourse', *Cartographic Journal*, **51** (1), 11–24.

Vujakovic, Peter (2018a), 'The Map as Spectacle', in Alexander Kent and Peter Vujakovic (eds) *The Routledge Handbook of Mapping and Cartography*, Abingdon, UK: Routledge, pp. 101–113.

Vujakovic, Peter (2018b), 'Cartography and the News', in Alexander Kent and Peter Vujakovic (eds) *The Routledge Handbook of Mapping and Cartography*, Abingdon, UK: Routledge, pp. 462–473.

Weigert, Hans W. (1941), 'Maps are Weapons', *Survey Graphic*, October, pp. 528–530.
Wood, Denis (1992), *The Power of Maps*, London: Routledge.
Wood, Justin (2005), '"How Green is My Valley?" Desktop Geographic Information Systems as a Community-Based Participatory Mapping Tool', *Area*, **37** (2), 159–170.

14 Closing thoughts and opening research pathways on geographies of slow violence

Shannon O'Lear

Introduction

This book project was off to a strong start and moving forward right on schedule ... and then COVID-19 hit, upending life and routine as we knew it. The delay in this project and many academic efforts pales in comparison to other effects of the pandemic and 2020 more broadly. The pandemic has brought to the world's attention multiple dimensions of harm most starkly measured in the zig-zagging charts of infection rates and a sobering death toll. Interwoven geographies of connections, movement, and boundaries have come clearly to light from the scale of international security protocols to the scale of the body and concerns with face masks and speech droplets.

COVID-19 has highlighted systemic forms of harm. This point is particularly true in the U.S. where these harms are so notable, because in many cases they could have been prevented and avoided. Deep and long-standing inequities based on ethnicity, skin color, gender, and age keep showing up on the more vulnerable ends of tables and graphs depicting the damaging impacts of COVID-19. Inequities are evident in the bleak realities of increased risk of exposure experienced by people fighting on the front lines in health care work. The inequities are also evident in the numbers of already economically disadvantaged and politically disenfranchised groups of people who have inadequate health care coverage, protections at work, or back-up options when schools go online and children stay home. Online schooling, furthermore, means that many children have lost access to consistent learning opportunities, valuable social interaction, and reliable meals and food security. Teachers have had less access to see when students might need intervention from everything from lapses in learning to dangerous home life situations. These are, unfortunately, familiar news headlines as the U.S. continues, as of this writing, to grapple with the virus, and they all demonstrate dimensions of slow violence that are being made more visible by the pandemic.

Simultaneously, there has been an eruption of attention to ongoing, systemic social injustice. There has been increased activity and visibility of the Black Lives Matter movement and other organizations and voices speaking up against physical violence, structural violence, and myriad practices of dispossession against black, Indigenous, and people of color. Intersectional identities involving gender, citizen status, environmental health, and economic opportunities complicate and add dimension to these social and political concerns. A range of voices point to the difference between what is and what could have been after many years of practices and policies that – deliberately or not – created injustices and impossible barriers for whole groups of people to thrive at their full potential.

Recap of chapters and looking ahead

The topic of this book is even more urgent at the time of publication. Perhaps it will offer useful insights into how making slow violence visible can amplify nearby work on various forms of inequity, injustice, and harm. As was pointed out in the introductory chapter, geographers are contributing a range of perspectives and methodologies in an effort to make otherwise latent and invisible harm more obvious. The chapters in this book have demonstrated a range of inquiry, methodology, and analysis to consider different ways that harm unfolds in a variety of spatial and social contexts, according to different temporalities, and through different social and structural processes.

The first few chapters of this volume highlight the importance of perspective and temporality. Through slow observation and an ethnographic focus on people's lived experiences of toxicity over time, Thom Davies offers an insight about what is actually invisible in these cases in his chapter, "Geography, time, and toxic pollution: slow observation in Louisiana." Although the physical or material toxicity is, itself, difficult to discern through single, snapshot observations, it is the people themselves, the people who are experiencing lived realities of environmental violence, as well as their stories and knowledge, who become invisible, overlooked, and discounted. This harm of a limited and unseeing perspective, Davies argues, is the true violence unfolding over time in toxic places.

In her chapter, "Rhythms of crises: slow violence temporalities at the intersection of landmines and natural hazards," Ruth Trumble has considered the simultaneous slow and fast temporalities of disaster and harm. The floodwaters that ravaged Serbia, Croatia, and Bosnia-Herzegovina created a disaster of stranded people, impassable roads, and suddenly evacuated populations in search of safety. Trumble's work examines, too, how the floodwaters also moved buried, unexploded ordnances from past conflicts as well as the signage that indicated their locations. Water and various forms of landmines moved across political boundaries and into areas at risk of landslides. The flooding posed an immediate danger, and the unknown location and stability of landmines added an unpredictable temporality to heightened danger. From her fieldwork, Trumble learned that gaining control over the landmines emerged as a priority in disaster response and evacuation efforts. Disruption caused by the flood accelerated the temporality of the mines, which had previously been understood as slow and relatively knowable, and made finding mines, marking them, and avoiding them an additional rhythm of the crisis.

John Paul Henry's chapter, "Complicating the role of sight: photographic methods and visibility in slow violence research," recognizes that forms of slow violence and toxic harm may be overlooked by an outside researcher. Therefore, it is all the more important to bring tools and methods that invite community members to reflect on the places, spaces, and processes through which they have experienced toxic harm in their everyday lives. The process of working with visual methods is collaborative and iterative. These methods not only help researchers to understand otherwise invisible forms of harm and damage in a lived, toxic landscape; they also help community members to articulate their experience and offer the potential to voice an alternative narrative about industrial practices that have become normalized and largely unchallenged.

In their work on Guatemala's Maya Biosphere Reserve, Jennifer A. Devine, Hannah L. Legatzke, Megan Butler, and Laura Aileen Sauls demonstrate how geospatial technologies can play into uneven colonial power dynamics. In the chapter, "Tourism development as slow violence: dispossession in Guatemala's Maya Biosphere Reserve," these authors demonstrate how different ways of seeing and different valuations of environmental features generate harm for some groups of people and the ecosystems on which they depend. In contrast to smaller-scale forestry practices and tourism activities that local communities offer and benefit from, outside actors are interested in expanding and commercializing tourism in the area. These actors are able to influence legal boundaries and support or challenge scientific categories, cultural identities, and the degree of visbility or invisibility of local activity through selective mapping and the use of LiDAR's multi-dimension visualization technology. These authors draw on their fieldwork and interviews to suggest three categories of slow violence associated with the global tourism industry: processes of land dispossession that may occur abruptly or incrementally, the commodification of culture and identity, and knowledge production about cultural landscapes. Attention to these processes can guide the prevention of these forms of slow violence in tourism and aim to decolonize processes of green neoliberalism.

Other chapters in this book also consider slow violence emerging from human-environment interactions. One of these is Daniel Abrahams's chapter, "From violent conflict to slow violence: climate change and post-conflict recovery in Karamoja, Uganda." In his interviews and conversations with Karamoja residents, Daniel Abrahams describes a trend in which people recall the direct violence of "those days" when cattle raids and armed violence dominated. Now, they recognize increased, complex vulnerabilities that emerge under conditions labeled as "peace." Although they feel safer from direct violence now, many people in this post-conflict context connect climate change impacts to an array of vulnerabilities ranging from failed agriculture to loss of identity, alcoholism, domestic abuse, and child labor. These compounding forms of difficulty and harm are not generally part of academic conversations that assume climate change will lead to conflict. However, they draw attention to multiple vulnerabilities that people without options for recourse will increasingly face as the physical systems of the planet become more unreliable due to impacts of a warming atmosphere.

Another look at human-environment interactions is offered by Kimberley Anh Thomas's focus on water infrastructure. Her chapter, "Enduring infrastructure," takes a relational approach to hydropower infrastructure that allows her to see the ongoing-ness of infrastructure. As a process, material infrastructure can involve creeping forms of violence and harm that unfold through phases of construction, operation and failure, and even decommission. In her discussion of water management infrastructure, she considers the temporality of harm, much of which unfolds slowly over time, the spatiality of harm, which is not necessarily co-located with the infrastructure itself, and the sociality of harm, which recognizes that some groups of people are harmed more than others through processes of infrastructure. In the examples throughout her chapter, Thomas demonstrates that slowly unfolding violence and suffering are not an exception to infrastructural processes, but are, in fact, integral to them.

Another set of chapters considers structural and social harm inflicted on particular groups of people. In her chapter, "Slow violence and its multiple implications for children," Sheridan Bartlett reminds us that conditions of environmental, economic, and social instability, while difficult and stressful for everyone experiencing them, tend to magnify and multiply negative impacts on children. As the least able to speak and act for or even defend themselves, children's experiences of myriad difficulties will likely compound and contribute to ongoing physical, psychological, and other forms of harm that endure and unfold over time. She discusses concerns about the longer term, undermining effects on whole societies when children know only difficult or harmful conditions in life. Children are not a minority group, and Bartlett encourages researchers to include age as an important dimension when studying social processes and generating policy suggestions.

In their chapter, "For Indigenous youth: towards caring and compassion, deconstructing the borderlands of reconciliation," Joseph P. Brewer II and Jay T. Johnson emphasize their focus on promoting inclusive and positive change rather than an intellectual engagement with the concept of slow violence. They describe in their chapter how Indigenous communities practice transformative justice that aims to restore order within a larger context of natural law rather than human-defined laws and systems of justice. They focus on Lakota young people living on reservations in North and South Dakota and describe both the roots and the ongoing practices of racism faced by these youths today. In telling their story and pointing out harm committed against a group of these youths, Brewer and Johnson acknowledge how the disregard of this harm, in effect, disregards the values of care and compassion that could otherwise build toward more equitable and inclusive justice. They argue in support of the decolonization of society across systems of politics, law, and science, and they encourage the difficult and uncomfortable dialogues that could, potentially, serve to overcome the slow erosion of communities and environment.

Michele E. Commercio considers another culturally embedded and systemic form of social harm in her chapter, "The infliction of slow violence on first wives in Kyrgyzstan." Polygyny, the taking of multiple wives, in Kyrgyzstan has re-emerged as a quasi-Islamist practice without a comprehensive, religious social structure to support it. At this time of economic and cultural difficulty, Commercio demonstrates how polygyny turns out to be emotionally and economically harmful to first wives, legally harmful to second wives, and devastating to all children involved who endure emotional instability and legal invisibility. The harm that this renewed practice inflicts on the social fabric of present-day society in Kyrgyzstan is multifaceted and will continue to do harm to family structures and children's identities for years to come.

The construction of public discourses and geographical imaginaries can be another means of doing harm by association. This is a key argument Aaron H. Gilbreath makes in his chapter, "When rednecks became meth heads: cultural violence, class anxiety, and the spatial imaginary." In his investigation of the cultural violence of "meth head" stereotypes, Gilbreath finds that media attention on certain places, such as the rural Midwest, served to overlay a limited, public understanding of methamphetamine with stereotypes of violent, poor, "rednecks." Familiar, widely distributed images of individuals' bodily decay in anti-meth campaigns reinforce the idea that meth users are rural, white people and hint at a precarious social position of white people in general. Merging stereotypes of poor, white people in low-income, rural places contributes to discourses that blame poor people for being poor. These public discourses are used to justify the continued marginalization, poverty, and denial of treatment for people in need. In this way, Gilbreath argues, cultural violence of public discourse can lead to forms of slow violence.

Legal definitions and practice, as discussed in the introductory chapter, establish and enact spatializations of violence often focused on particular groups of people. In their chapter, "The slow violence of law and order: governing through crime," Samuel Henkin and Kelly Overstreet examine how the making and implementing of law can shape the slow violence of (in)justice. Despite the aura of neutrality and impartiality, there is a lack of transparency in how prosecutors decide who, how, and when to charge people for alleged crimes. Through this unchecked power, prosecutors wield significant influence in determining the life chances of individuals. The authors highlight the wrongful conviction and imprisonment of six African American youths by a high-profile prosecutor known to adhere to 'law and order' practices. Although the convictions were all vacated years later, after the defendants had all served multiple years in various prison facilities, the prosecutor never admitted to wrongdoing. It is practices such as this that perpetuate slow violence and the (re)production of various forms of discrimination and unjust social and spatial relations.

Finally, Peter Vujakovic's chapter on dark cartographies delivers a uniquely geographic perspective on how violence is communicated, stabilized, and practiced through cartographic work. This chapter also speaks to how geographic knowledge is mobilized to create worlds. In the chapter, "Dark cartographies: mapping slow violence," Vujakovic draws on historical and contemporary examples to demonstrate ways in which maps and other geospatial technologies, such as drones, enable the projection of power by elites. He considers how the naming and un-naming of places standardizes and normalizes one sense of place over others, reflecting practices of possession, dispossession, subjugation, and cultural violence. News maps in mainstream media play a particularly important role in educating the public and shaping public opinion and spatial (mis)understanding; accuracy is often sacrificed for dramatic impact when maps become tools for 'infotainment' and examples of 'chartjunk.' Vujakovic's chapter also looks at ways in which critical cartography in both theory and practice draws attention to spaces of injustice and offers an alternative possibility. He describes several examples of 'oughtness' mapping and sociological mapping over time that have depicted otherwise invisible forms of harm and cultural violence as a way to challenge the status quo. Mapping and cartographic practice, he argues, are vital if we are to make slow violence visible.

In all, the chapters in this volume demonstrate a number of ways in which geographers and scholars in nearby disciplines can work to make slow violence visible and otherwise bring attention to forms of harm and injustice that are embedded in political, social, and economic structures and intertwined with legal, policy, cultural, scientific, and technological practices. It matters who is looking and who is communicating, who is defining categories, and who is drawing the map. Expanding research design to consider slow violence, structural harm, and other forms of indirect neglect renders research more robust and the findings of that research more complete in the story it can tell. Many of the chapters in this volume emphasize, however, that it is not wholly in the researcher's hands (or eyes) to see slow violence. Collaborative, qualitative, and ground-truthed projects allow researchers to see through and beyond numbers and metrics that limit our view to what we already know. To see the harm that we cannot readily see, and to understand the causes and impacts of that harm, are the core purpose of paying attention to slow violence in research.

The endeavor of research is to find things we cannot readily see, to identify connections that are not obvious, and to understand places, processes, and patterns in new and comprehensive ways. Centering research on questions aimed at disentangling the construction and experiences of invisible, latent, indirect violence is challenging, and the chapters in this book offer ways to approach this kind of work. It is also worthwhile to keep slow violence in mind when designing research and to leave space within the research framework to allow for the consideration of this kind of harm should it emerge during the research process. Unfortunately, there are many ways for that to happen. This book, hopefully, provides useful ways to recognize slow violence and otherwise hard-to-see forms of harm so that they may be brought into research and made visible through our collaborative investigations of the world.

Index

Printed and bound by CPI Group (UK) Ltd, Croydon, CR0 4YY

06/02/2023

03188147-0002